Infrastructural Ecologies for Fouché, Haiti: Multipurpose, Integrated and Synergistic Systems

- OVERVIEW
  - Scales & Systems 3
  - Area-wide Initiatives 3
  - Modular Scale Initiatives 6
  - Flows 6

- PROJECT COMPONENTS
  - Biodigesters 8
  - Agriculture / Permaculture 13
  - Water Resources 17
  - Concrete Production Site 22
  - Waste Management 25
  - Heat and Power – Community Hub and Eco-Industrial Park 29
  - Community Hub 33

- Overall Program Cost 36

- Attributions 37

- Acknowledgements

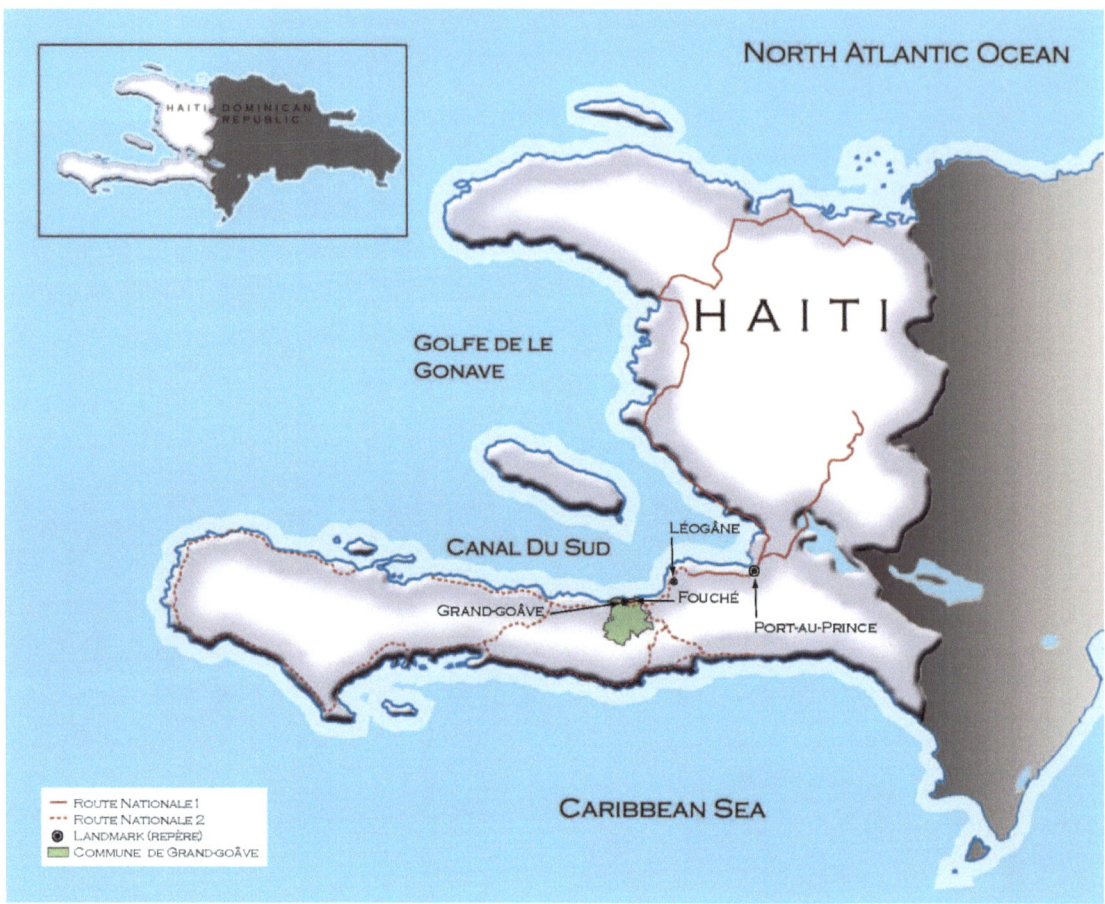

Map of Fouché and Surroundings

Overview

How might a small Haitian village (estimated population: 1,400) lacking basic power, sanitation, waste collection and with a limited water supply, achieve a measure of infrastructural self-sufficiency? What means might be developed to provide these basic services relying primarily on local natural resources? Could these services create local jobs? This was the problem statement set out for the farming settlement of Fouché, a hamlet spread along a main road running through Les Palmes district in the Ouest region of Haiti. Fouché lies on the coast about 10 miles west from Léogâne, the epicenter of the devastating January 12, 2010 earthquake.

Much technical and economic development assistance in Haiti today is rendered on a sector-by-sector basis, with support typically compartmentalized according to the specialties of housing, agriculture, reforestation, sanitation, or energy, etc. *Infrastructural Ecologies for Fouché, Haiti* adopts a multi-objective, holistic design approach reliant on an integrated planning process. Exchanges across the sectors of agriculture, water, energy and waste "close the loops" of energy and resource flows, i.e. waste from one sector supports another. While some of the recommendations in this report can be implemented independently, this collective approach potentially provides the highest return on investment through job creation, as well as significant social and environmental benefits.

The project illustrated here has three objectives. The first is to offer Les Palmes regional and local government and their NGO affiliates a vision for redeveloping their rural settlements for improved self-reliance, accomplished through development of basic critical infrastructure services, the management of which creates local employment. The second is to offer reliable, low-tech alternatives to the carbon-intensive infrastructure promoted by the industrialized world. The project's whole-systems solution takes advantage of cycling local resources. The proposed systems and technologies are phased to be built, maintained and operated largely by trained local labor. The third objective is to offer the concepts and constructs of "infrastructural ecology" as a template for redevelopment in other rural regions across Haiti.

Scales & Systems

Infrastructural Ecologies is a multi-objective, integrated redevelopment scheme envisioned for the rural, tropical, and local conditions of Fouché, Haiti that is widely applicable to comparable agrarian areas. Closed-loop systems are considered at two scales:

A. Area-wide (broad, community initiatives)
B. Modular (concentrated, housing-cluster initiatives)

Along these scales, three systems aim to address adversities afflicting Haiti at large, as well as those local to Fouché:

1. The Bioremediation & Food System addresses erosion and topsoil loss from deforestation; aquifer depletion, food and nutritional shortages, and limited access to potable water;

2. The Waste & Energy System addresses the burning of imported carbon fuels for power and locally harvested wood for charcoal; unsanitary conditions caused by disorganized solid waste flows, groundwater contamination, and lack of local electricity to support lighting and communication;

3. The Community System addresses the lack of sustainable commerce, under-employment, the high costs and low quality of concrete for construction, and lack of communal and global connectivity.

Area-wide Initiatives

- Flooding reduction and partial diversion of waterways into a network of ponds; the addition of an upland vetiver grass sediment barrier, as well as other flooding buffers to help recharge aquifers and revitalize the soil

- Some 19 local, edible plant species systematically grouped to maximize the nutrient-exchange benefits of complementary root structures, and canopy levels, to create a diverse food source for the population of Fouché

- Agricultural waste collected for processing in a community biodigester, yielding biogas for cooking in communal and modular kitchens, and fertilizer for enrichment of agricultural land

- Excess harvested food is channeled to a restaurant and food dryer, a new complex, the Community Hub, which serves as a source of communal income and jobs

- The Community Hub also unites and connects Fouché inhabitants by incorporating a town clinic, computer laboratory, and closed-loop communal laundry center

- Expansion of an existing gravel depository into a small eco-industrial park housing a new commercial concrete block operation, a recycling and disposal center for Fouché's inorganic waste through a dedicated landfill

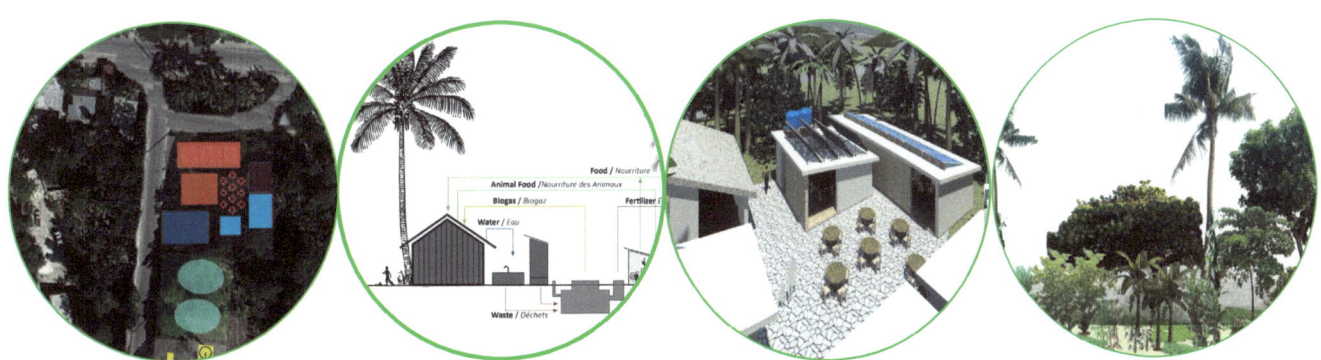

The Community Hub will provide computer access, dried fruit production, a laundry facility, restaurant, clinic, and employment for the town. Biodigester proposals will reduce organic waste pollution and provide each family with biogas for cooking. An Eco-Industrial Park will be a source of income, employment, and quality concrete building blocks. The agricultural proposals seek to remediate the land, reverse the effects of erosion, increase farmland productivity, and provide a source of food and employment for the community.

Infrastructural Ecologies for Fouché, Haiti: Multipurpose, Integrated and Synergistic Systems

Fouché Existing Conditions

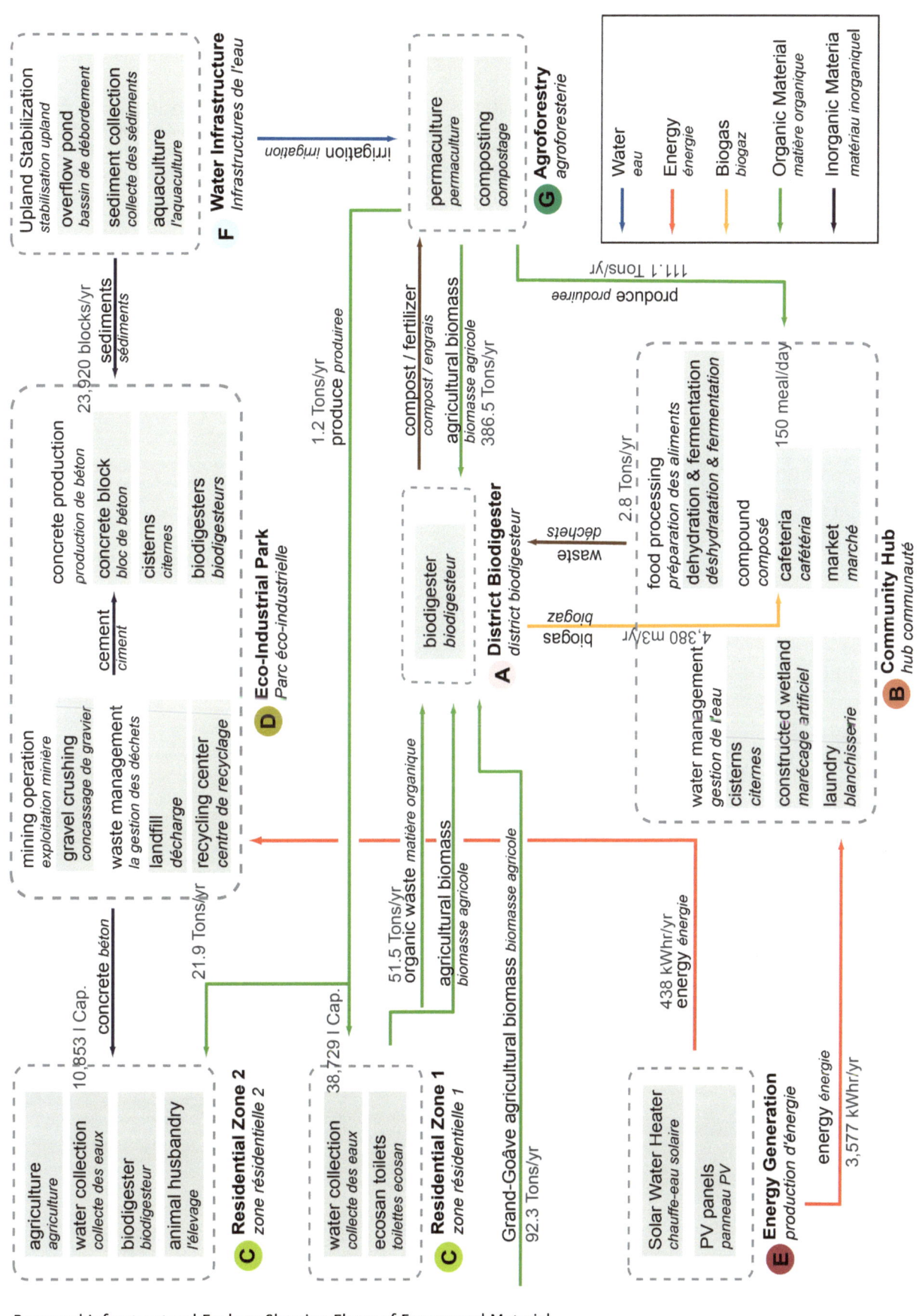

Proposed Infrastructural Ecology Showing Flows of Energy and Material

Infrastructural Ecologies for Fouché, Haiti: Multipurpose, Integrated and Synergistic Systems

Modular Scale Initiatives

- A garden, water catchment pond, and animal sheds comprised the key elements of a single, closed-loop system for each 20-house cluster (roughly four groupings of 5 existing or new houses), providing a source of fertilizer and irrigation for the garden, which supplements communal agriculture yields

- Each 5-house cluster is equipped with three composting toilets, a rainwater catchment cistern, and small biodigester; collectively, these elements serve to provide basic sanitation, increase water supply, and process the cluster's domestic waste into a shared cooking fuel.

Flows

The organizing principle of these self-sustaining systems is the cyclical flow of food, waste, biogas, fertilizer, and utilization of local materials for construction. Below is a high-level depiction of these transfers.

The diagram on page 5 shows the closed-loop sequence of energy, water and waste transfers that collectively support basic community services.

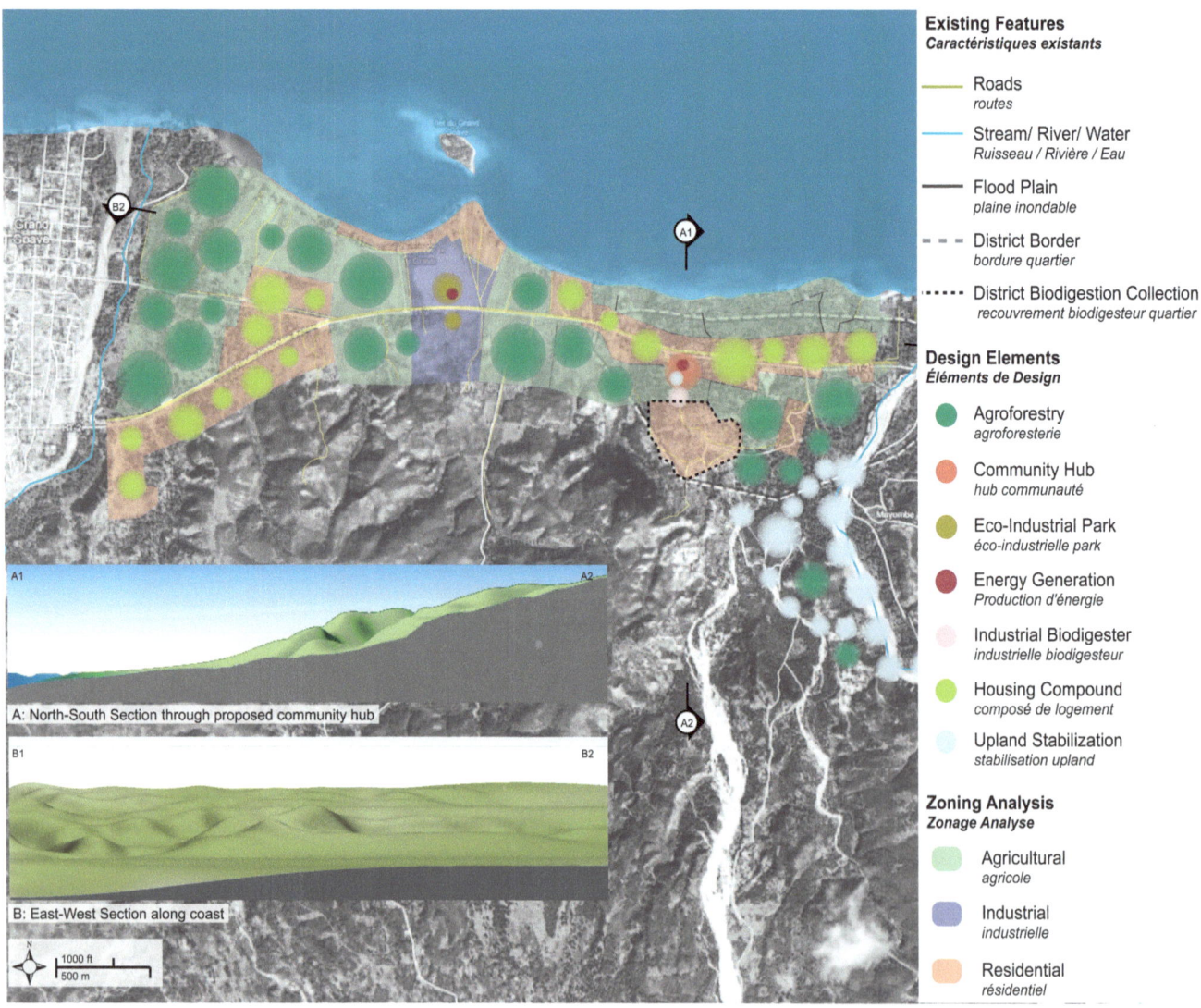

Fouché Proposed Conditions

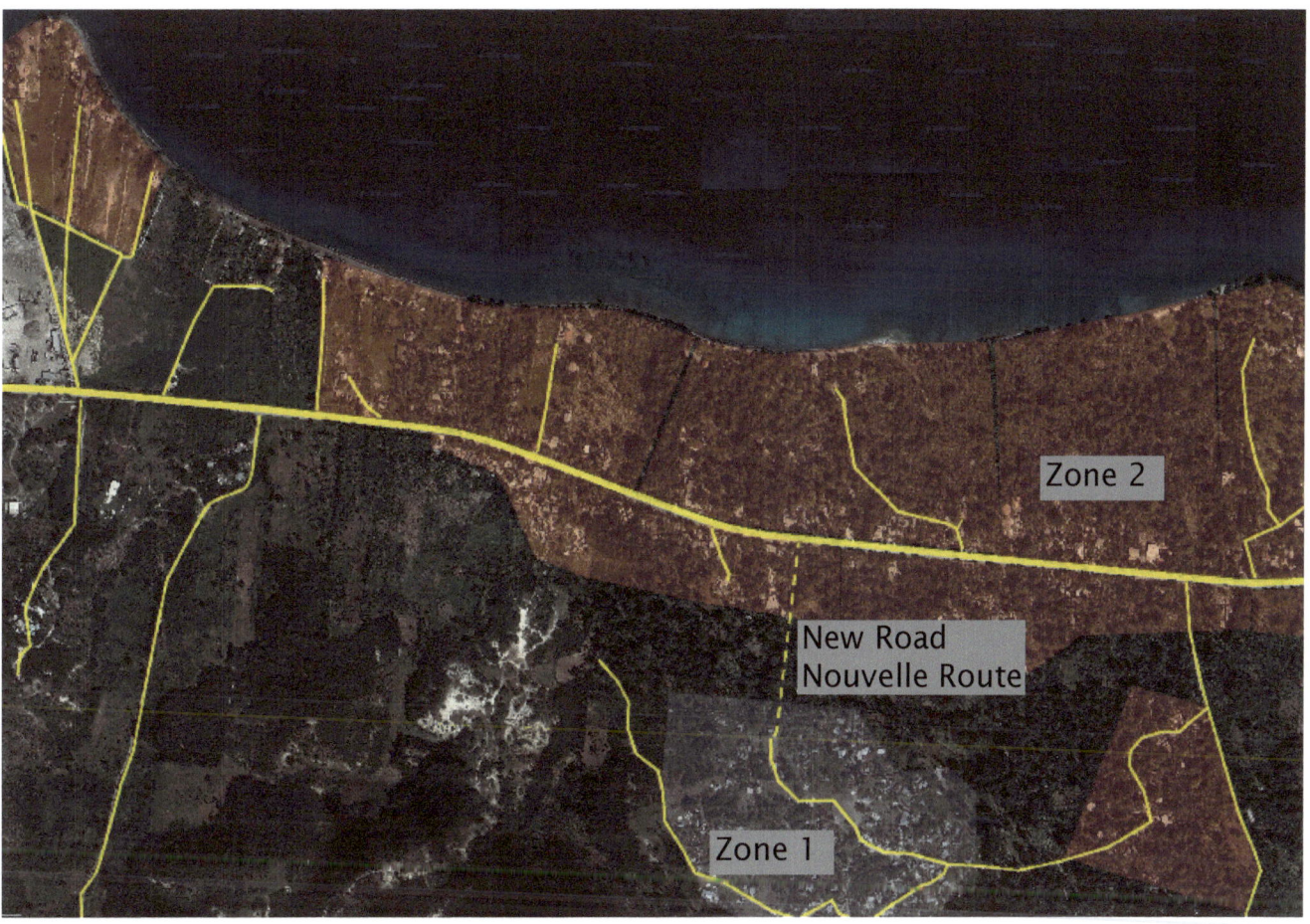

Fouché Area Zoning

Zone 1
- 140,928 m2
- 89 buildings
- 400 inhabitants

Zone 2
- 220 buildings
- 990 inhabitants

Biodigesters

Proposed Community Hub Site Plan

Problem statement

In Fouché, Haiti, where approximately 56% of the waste stream (agricultural, domestic) is organic material, a great proportion that is not composted or fed to animals is disposed of improperly (discarded) or underutilized (burned, when it could be composted). Organic waste turned to useful biogas energy could offset charcoal use for cooking, which causes deforestation and subsequent erosion, compromising arable land by loss of nutrients.

Proposed Solution

The application of biodigesters addresses these problems. Biodigestion processes the above-described combined organic waste into useful biogas with fertilizer as by-product. A biodigester provides the anaerobic conditions that foster chemical reactions (hydrolysis, fermentation, acetogenesis, and methanogenesis) that rely upon bacteria: fermentative, acetogenic, and methanogenic. Fouche's climate supports an ambient temperature adequate for these processes to occur without much added heat. The end product, biogas, can be used as cooking gas with conventional burners, while the fertilizer slurry generated can help remediate the depleted soil nutrients. Biodigesters have been successfully used all over the world in similar climates, including China, India, Brazil and Central America. Successful cases of small, inexpensive biodigesters abound in the Central American countries of Costa Rica and Guatemala.

In Fouché, biodigesters are divided into two major categories by size: small, local biodigesters and a mid-sized, district-scale biodigester. The small biodigester serves the Zone 2 household clusters, while the mid-sized biodigester, located in the community hub, processes waste collected from agricultural areas and Zone 1 households.

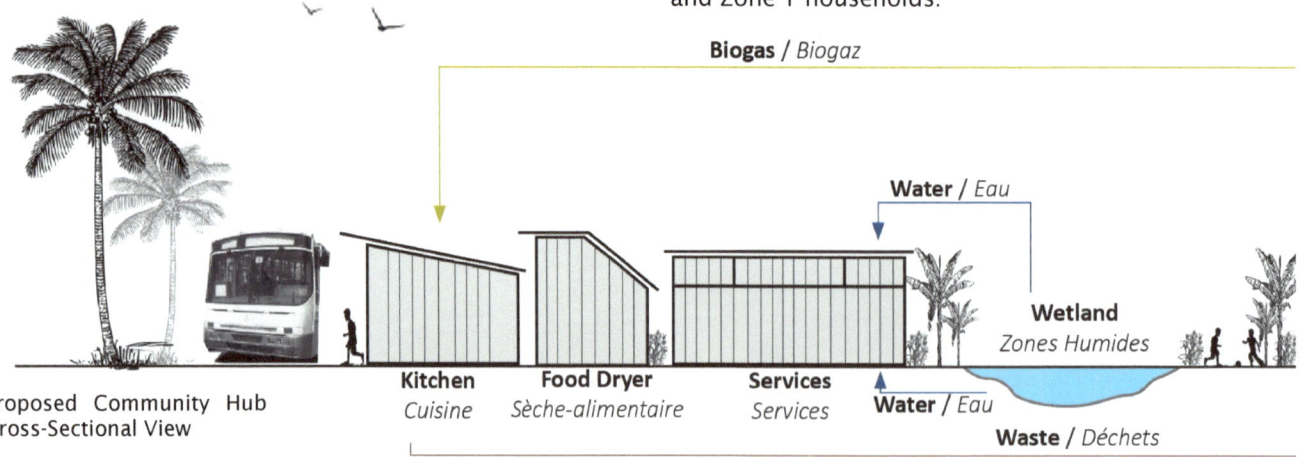

Proposed Community Hub Cross-Sectional View

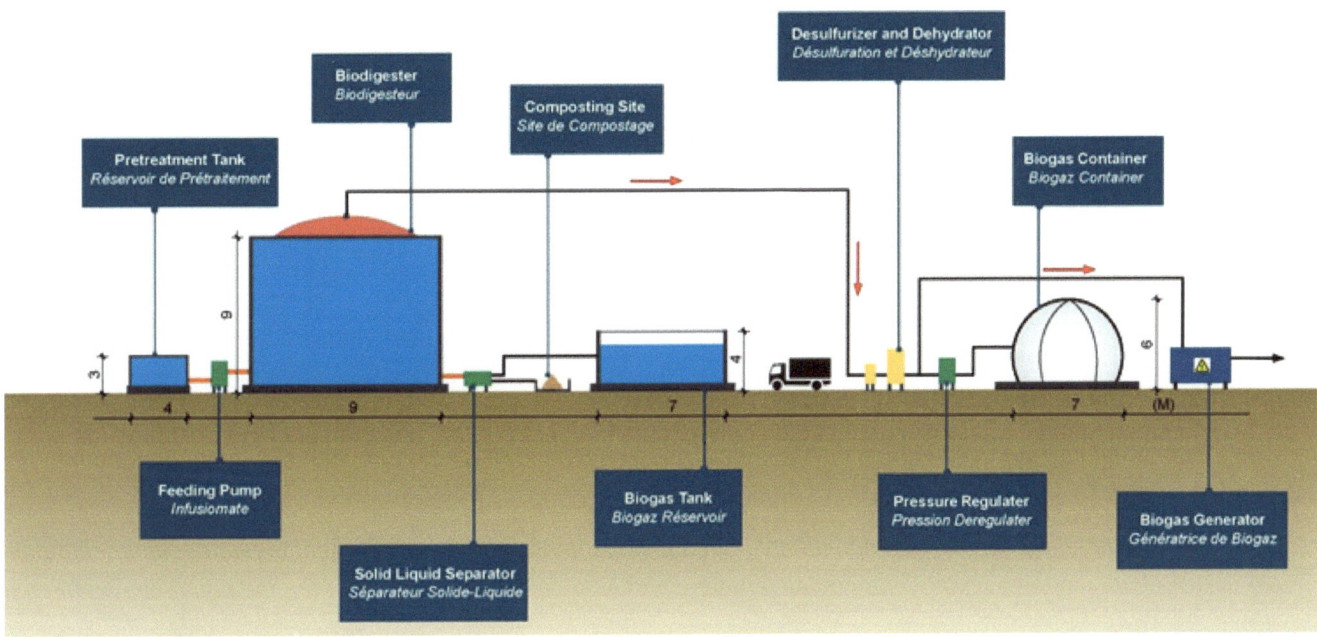

Typical Community-Scale Biodigester

The gas generated from the former, can provide a little over an hour of cooking time per household, per day. Additional Zone 1 gas adds another 1.1 hours per household per day if distributed evenly. Excess can be sold for profit. Biogas generated from the district biodigester can power 4 burners for 8 hours a day year round in the community hub restaurant. Fertilizer goes back to the agricultural zone.

Fouché households own goats, chickens, and tend local farms. Gardens for the 20-household clusters in Zone 2 supply crop waste to be combined with human and animal waste. Each biodigester serves 5 houses, is connected to 3 toilets, and is situated next to an animal pen for easy transport of wastes. Organic waste trucked from Grand-Goâve, combined with the semi-annual harvest waste, in addition to routine domestic waste should support the continuous fueling requirements of the district biodigester.

Challenges

Biodigesters must be carefully sealed to prevent gas leakage as it includes CH_4 (methane, which is flammable), as well as CO_2, H_2, N_2, and H_2S. Construction and operational training of local users is critical. Both the small and middle-scale biodigesters require constant loading and correct balancing of the organic waste mix. If the flow stops or becomes unbalanced the biodigesters will stop working, and restarting is a difficult process, therefore Fouché must have stable sources of different kinds of organic waste.

Proposed Phasing

Phase 1 (3 months)
Small biodigester: Training local users in the construction of local small-scale biodigesters, the start of construction, and preparations for waste collection;
Large biodigester: Site clearing and design for the district biodigester

Phase 2 (~5 months)
Small biodigester: Start of operations;
Large biodigester: civil work, construction of biogas containers, purchase of necessary equipment

Phase 3 (4 months)
Large biodigester: (constructed by experienced contractors).Installation of equipment, electricity and gas devices, testing, and training of workers; finish construction work and begin use

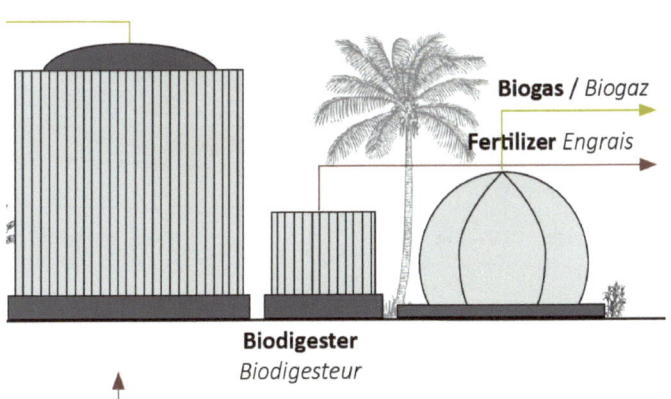

Biogas / *Biogaz*
Fertilizer *Engrais*

Biodigester
Biodigesteur

Infrastructural Ecologies for Fouché, Haiti: Multipurpose, Integrated and Synergistic Systems

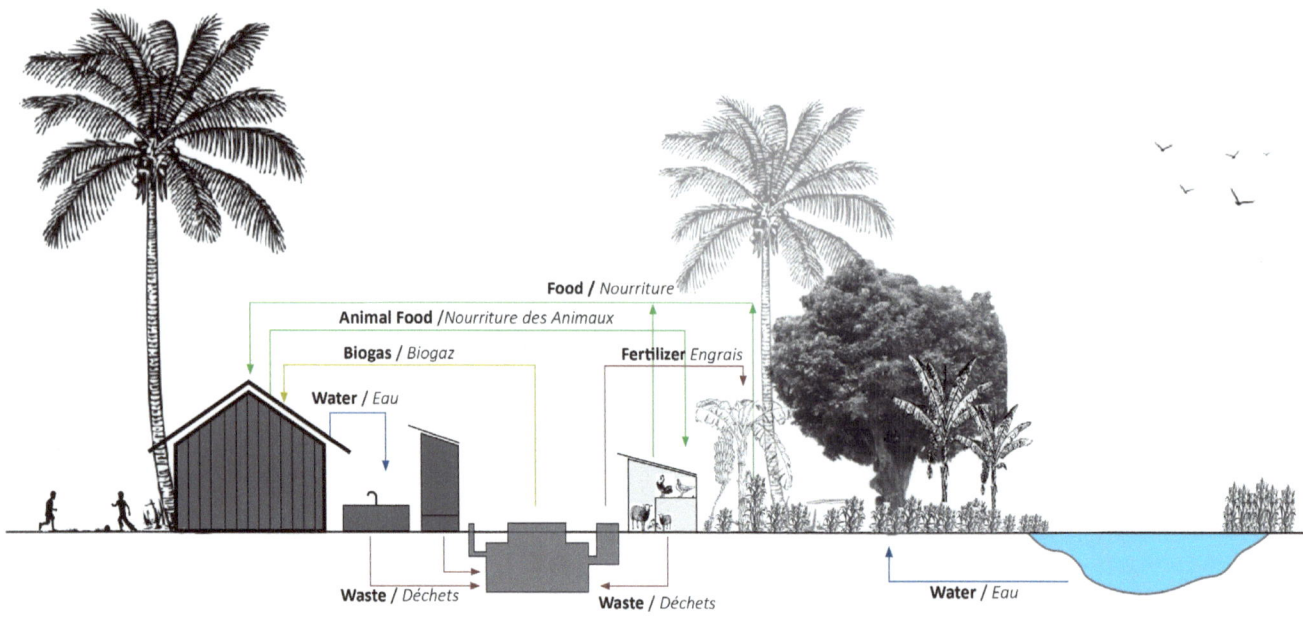

Proposed Typical 5-House Cluster Cross-Sectional View

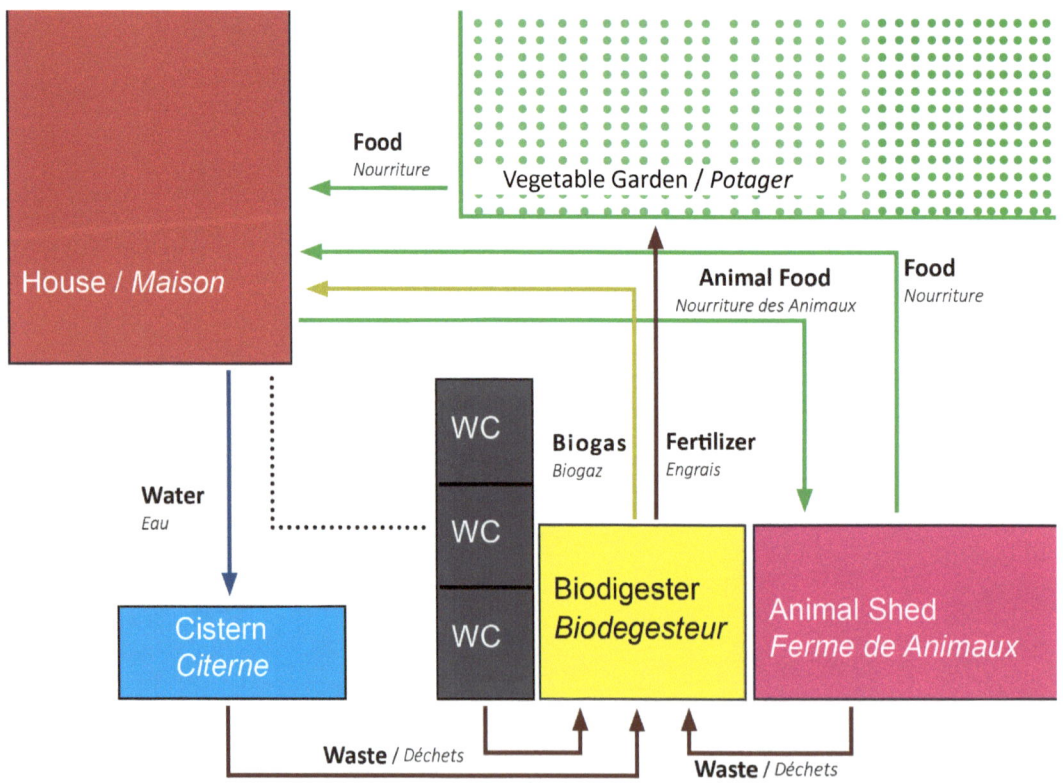

Prototype of 5-House Cluster Resource Flows

Biodigesters

Proposed 5-House Cluster Site Plan

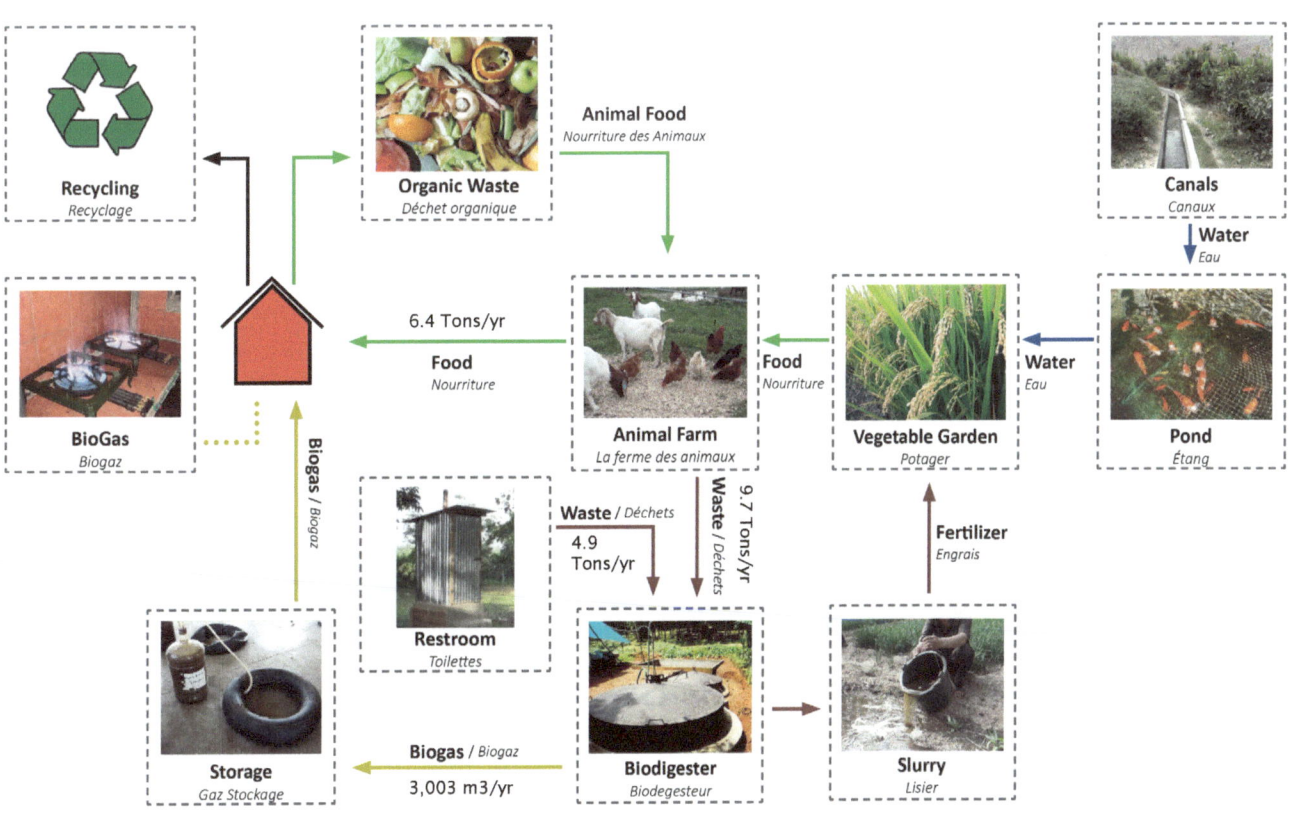

Proposed Biodigester System Flows for each 5-house cluster

11

Infrastructural Ecologies for Fouché, Haiti: Multipurpose, Integrated and Synergistic Systems

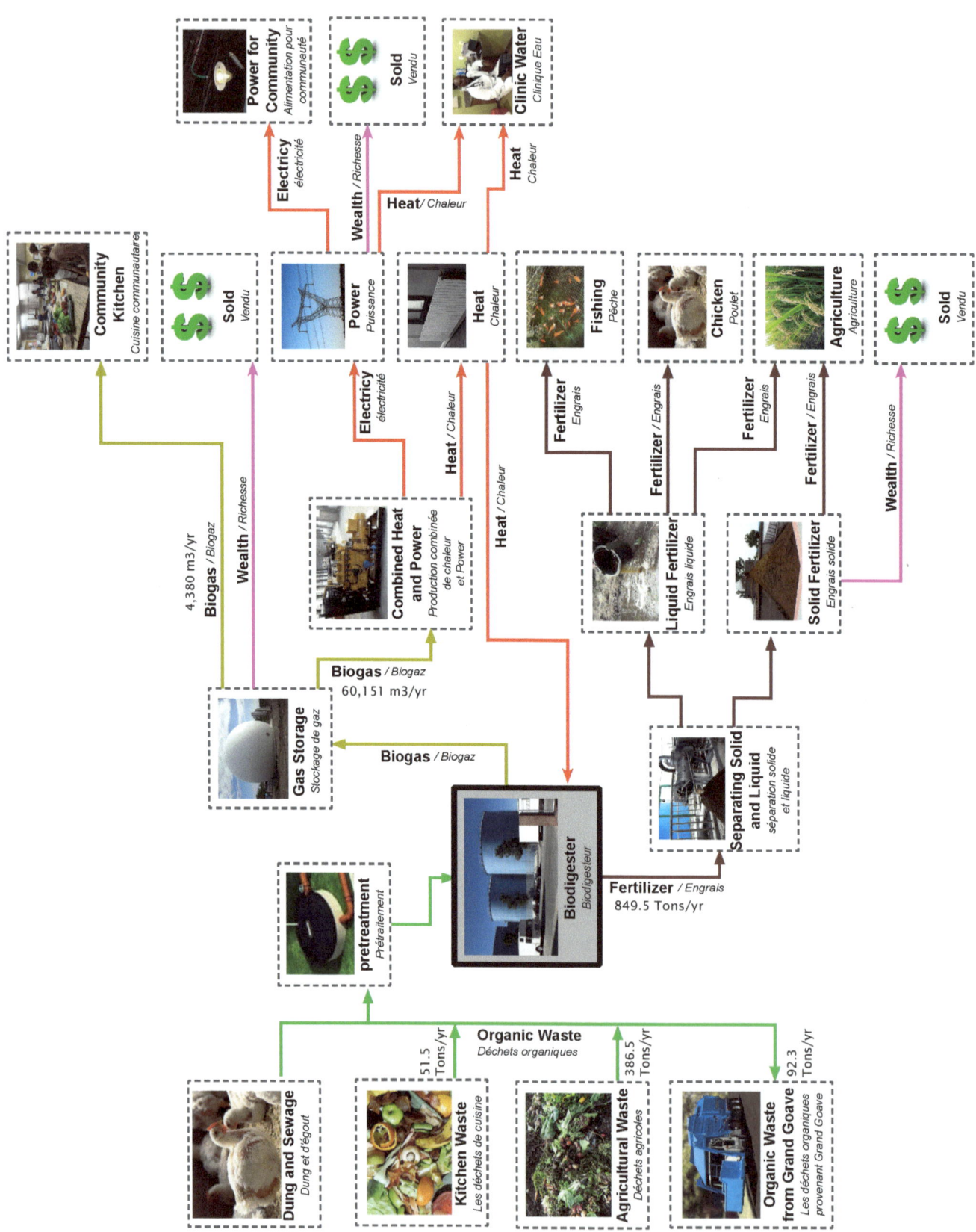

Proposed Area-wide Community Biodigester System Flows

Agriculture/Permaculture

Problem Statement

Haiti's widespread deforestation has caused routine flooding and major loss of fertile top soils vital for ecological diversity and agriculture. Further, instability of land ownership has been a major contributor to deforestation. Permacultural practices ("permanent agriculture") can help remediate soils and prevent further erosion, improving overall land productivity.

Proposed Solution

Existing land in Fouché may be categorized as agricultural, uncultivated, or arid. Agroforestry and intercropping, applied through the introduction of an "edible rainforest garden" can boost agricultural

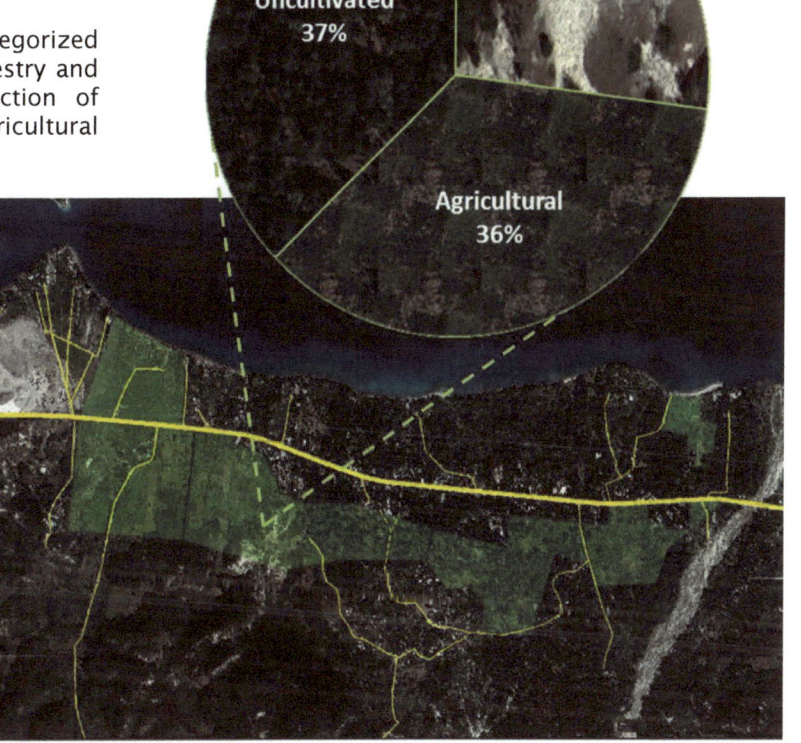

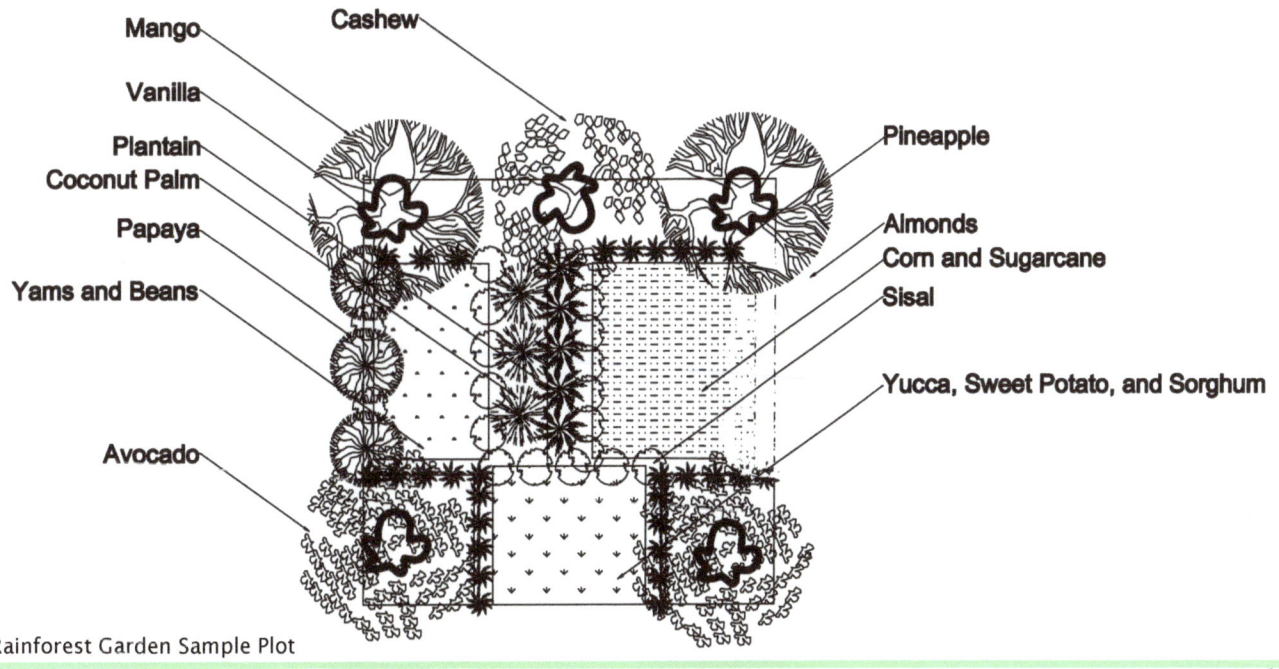

Fouché Agricultural Zone, Existing Conditions

Rainforest Garden Sample Plot

Proposed Edible Rainforest Garden

yields in both cultivated and uncultivated lands. It can regenerate soil, thus bringing food security and a measure of economic stability to the rural settlement of Fouché.

Given the areas' proclivity to host a multitude of edible vegetation, these rainforest gardens will consist of plants native to Haiti, strategically placed for mutual benefits and soil strengthening. The twenty different crops proposed in the design not only provide greater overall yield than monoculture plots per unit area, they interact to reduce pests, increase nutrients, improve the local diet, with surplus crops dried for retail sales. Intercropping of different species and plants of different heights improves shade, increases water retention and promotes symbiotic root interactions. Overall, the rainforest garden remediates depleted soil conditions while creating a diversified local food source. Shared gardens are recommended for each 20-unit housing cluster in rural Zone 2 as additional food sources as well as feed for husbanded animals. Estimated combined yields, given 2 harvests per year and rotating farm lands, provide sufficient food for 1.25 kilograms of produce per person, twice a day, annually in addition to 1.5 kilograms of food per meal, for 150 meals a day for the community hub restaurant. Moreover, 29 tons of food per year can be processed through the community food drier.

Challenges

Reliable sources of water are necessary for this large contingency of plant biomass. A macro-catchment pond diverted from the River Canot can store and distribute irrigation to agricultural plots in Fouché. Zone 2 gardens will be irrigated from small stormwater storage ponds. These may require solar-powered water-circulation devices to reduce harmful bacteria from stagnant water. Otherwise, these pools can be used only for shouldering the rainy seasons. Finally, training and education is required for both agricultural land and Zone 2 gardens to ensure the proliferation of permacultural practices.

Proposed Phasing

Permaculture designs are readily implemented due to the prevalence of the name crop species, with the exception of vanilla bean. Plots of mixed crops that require the most sunlight should be planted first, while sheltering some of saplings that will comprise the canopy layer. Canopy trees, as they mature, stabilize soils through the root structures. Sub-canopy trees, shrubs, and the vanilla bean vine can be planted last. With the edible rainforest garden in place, gardens and water catchment pools for the 20-household Zone 2 clusters can be established, with the total implementation of the plan can be estimated at a little less than 1 year.

Agriculture/Permaculture

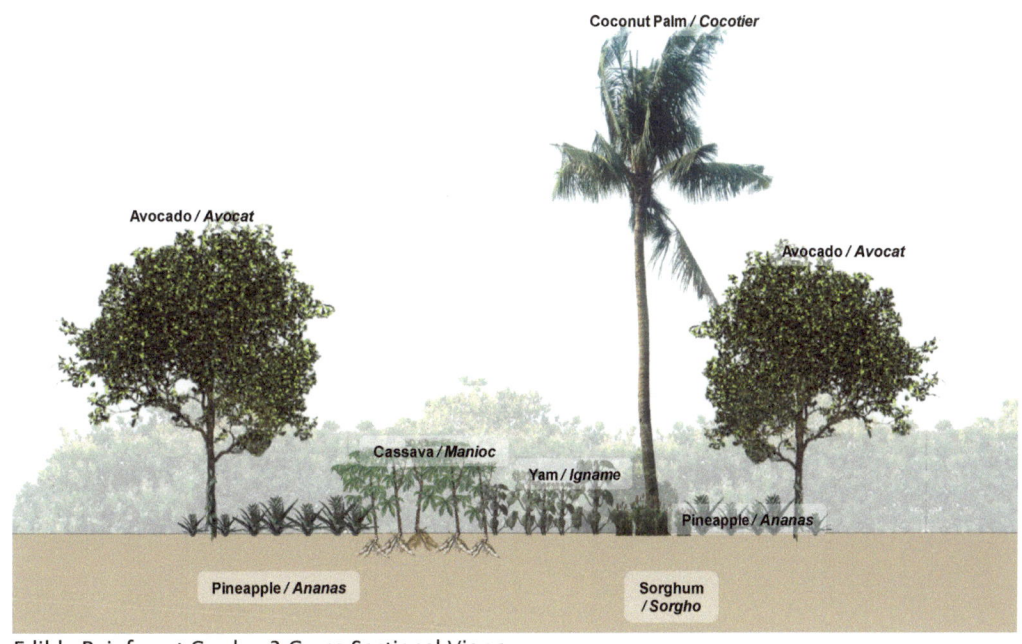

Edible Rainforest Garden 3 Cross-Sectional Views

Infrastructural Ecologies for Fouché, Haiti: Multipurpose, Integrated and Synergistic Systems

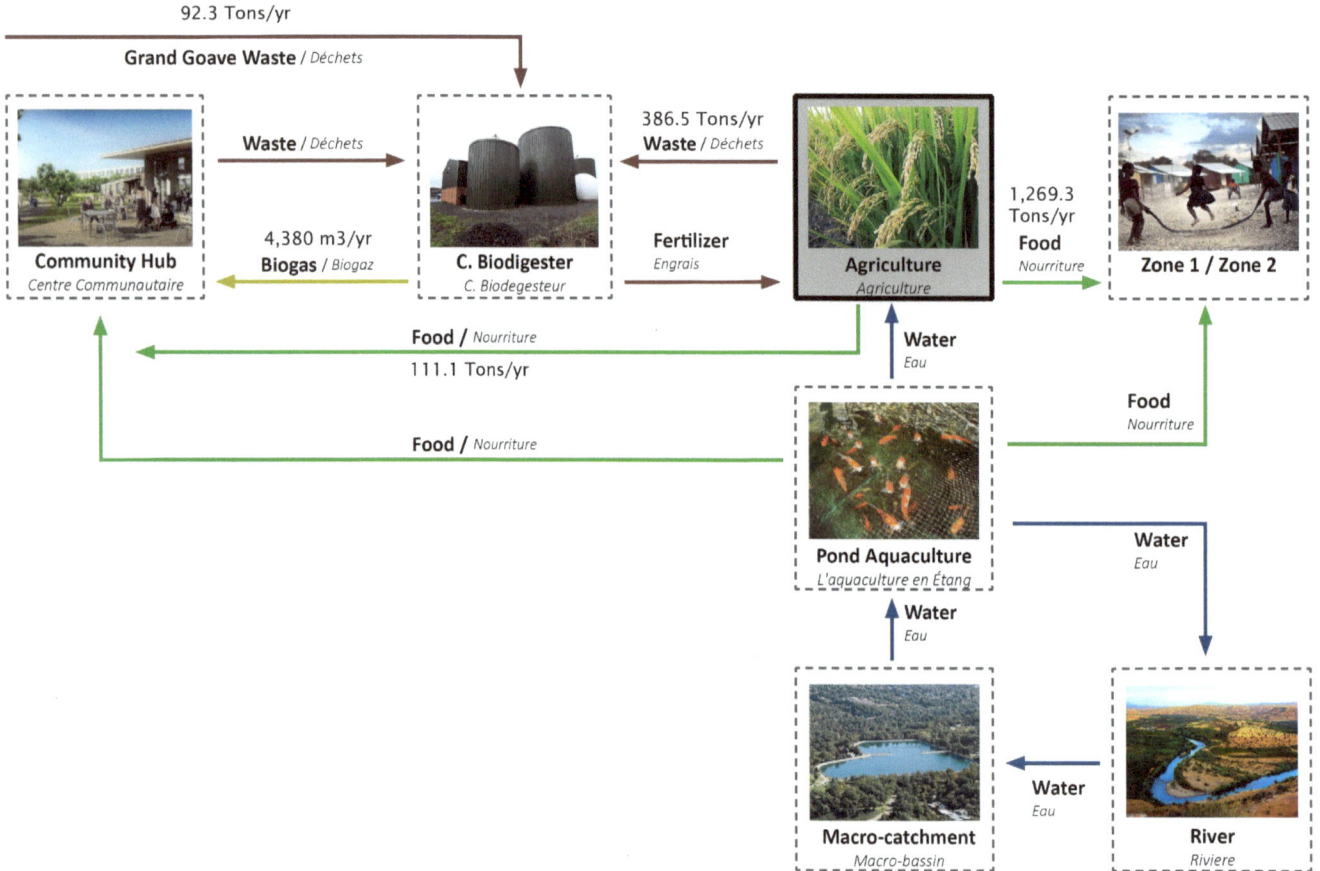

Proposed Agricultural Zone System Flows

Water Resources

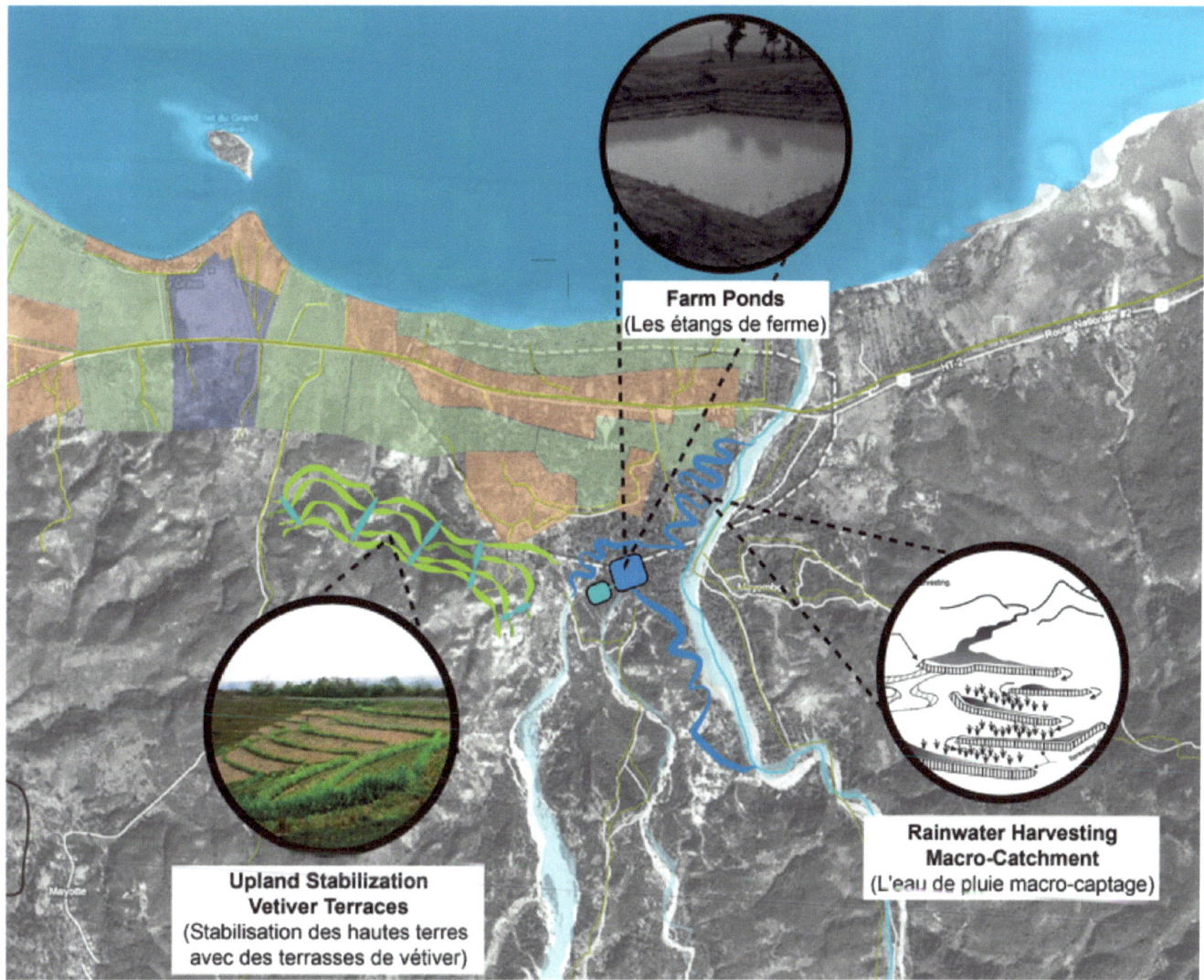

Fouché Proposed Water Resources

Problem statement

A critical challenge in Fouché, as in much of the rest of Haiti as a nation facing water scarcity, is procurement of sufficient water to serve both potable and non-potable needs. Three wells, only one potable, supply the settlement. Neither water supply piping nor water storage tanks exists. An interrelated crisis is the lack of sanitation infrastructure and present unsanitary washing practices, particularly in rural areas like Fouché. The disruption of the hydrologic cycle in Haiti due to climate change (severity of storms), deforestation and loss of plant and soil ecosystems that retain surface- and groundwater is a major problem.

Proposed Solution

This proposal takes advantage of the geographical, topographical and climatic conditions in the settlement to create water source diversity and beneficial relationships between water and the agricultural sectors. It additionally focuses on watershed recovery and conservation, requiring the active participation of the community. Rooftop rain water harvesting systems (RWHS) are to be implemented in each housing cluster for domestic water supply, reducing pressure on the single potable town well. Storage will be accommodated through the construction of shared cisterns. The Ecological Sanitation (EcoSan) initiative successfully piloted elsewhere in Haiti creates a closed loop between sanitation and agriculture, reducing groundwater pollution. Dual dehydration diversion toilets will be placed in each housing cluster. These simple constructions dehydrate excrement, collect urine, and sanitize it for use as a high level nutrient for improving agricultural productivity. At a macro level, existing rivers will be diverted into a network of ponds and canals for use in agriculture and aquaculture. An upland stabilization program will utilize tiers of vetiver grass for catchment of sediments, revitalizing organic soils and ground water recharge. These macro scale strategies provide the town with a flooding buffer during the rainy season, avoiding economic loses from crops and housing damage.

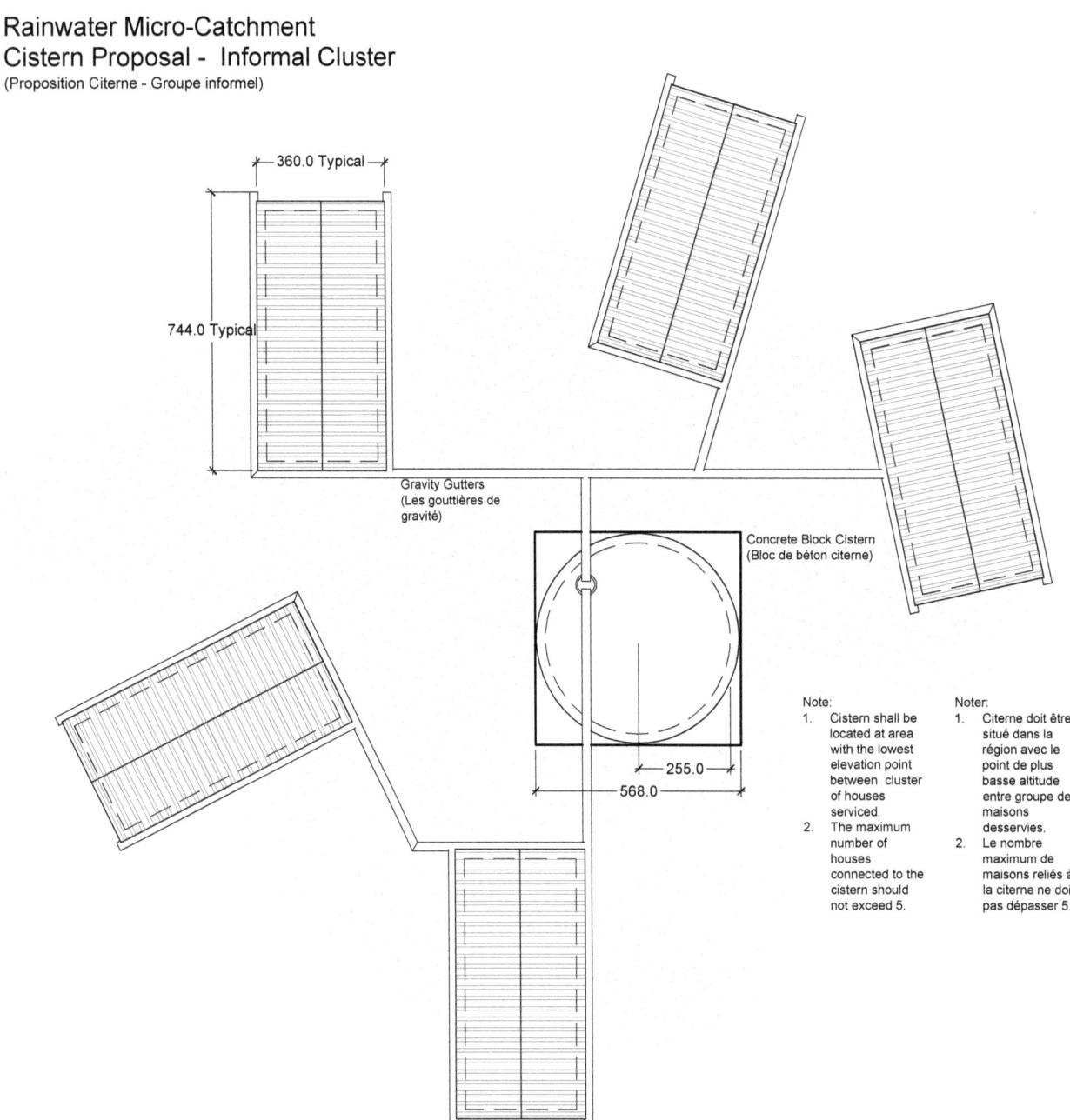

Rainwater Micro-Catchment
Cistern Proposal - Informal Cluster
(Proposition Citerne - Groupe informel)

Challenges

Community engagement and education will be vital to the success of this program as current cultural practices and personal habits pertaining to clean water supply, sanitation, and drainage must be improved.

Proposed Phasing

The first imperative will be to construct eco-sanitation systems (composting toilets or connections to the biodigester). Another priority is the upland water management diversion, linked with the agricultural development and followed by the RWHS implementation.

Water Resources

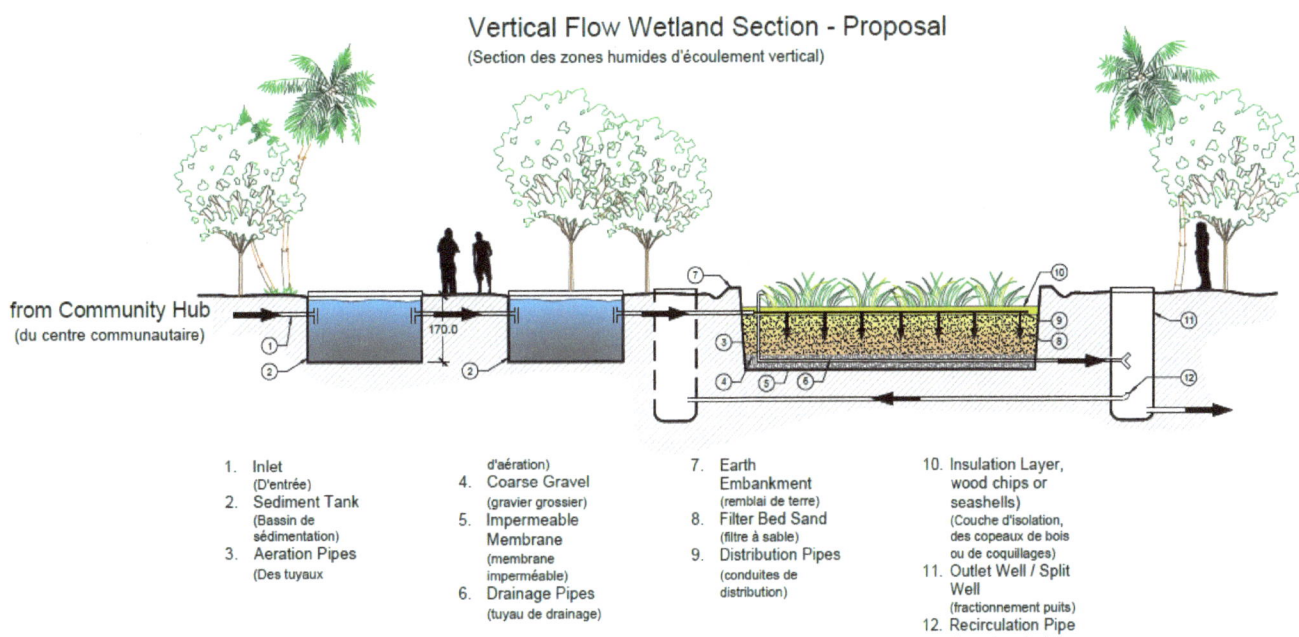

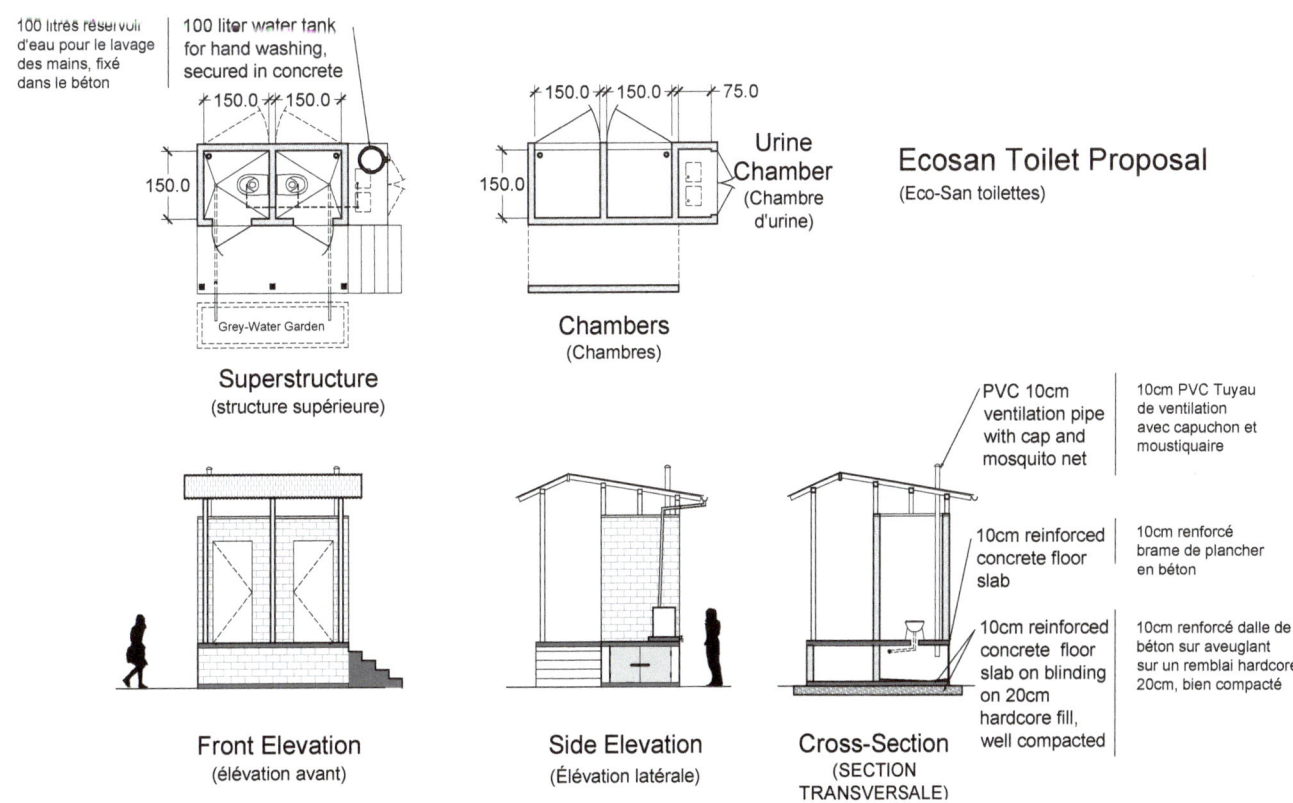

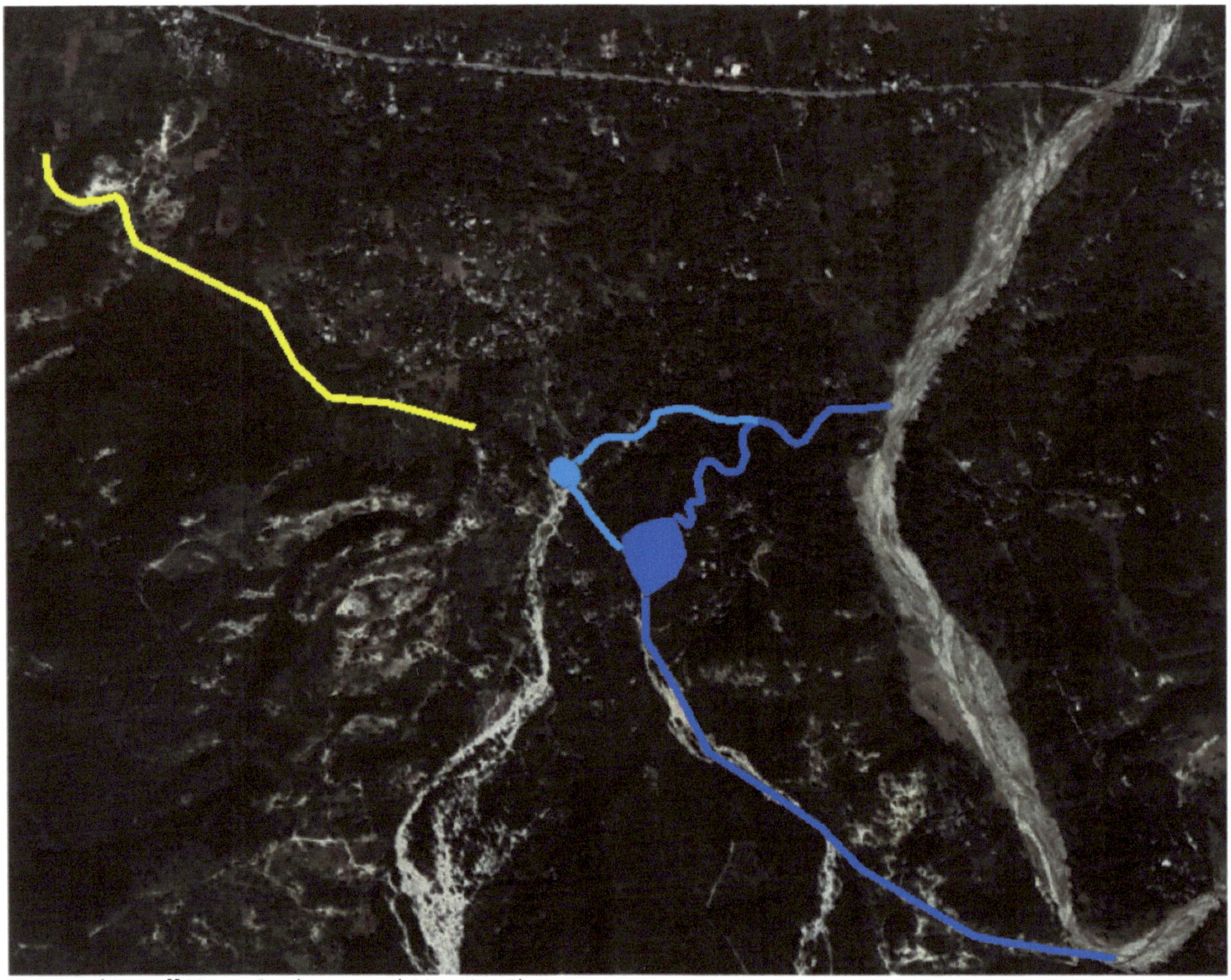

Infrastructural Ecologies for Fouché, Haiti: Multipurpose, Integrated and Synergistic Systems

Proposed Runoff Macro-Catchment and Diversion Plan

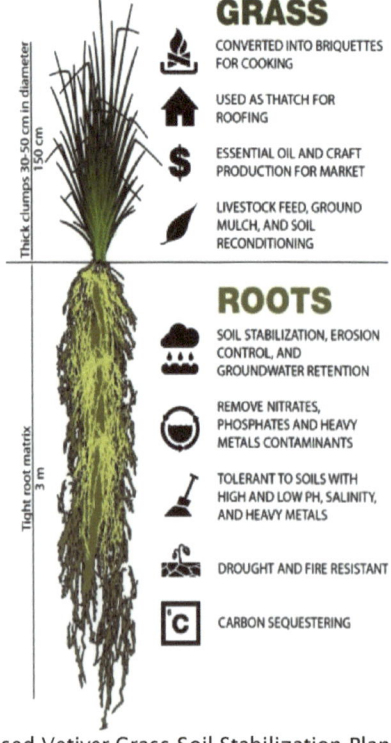

GRASS
- CONVERTED INTO BRIQUETTES FOR COOKING
- USED AS THATCH FOR ROOFING
- ESSENTIAL OIL AND CRAFT PRODUCTION FOR MARKET
- LIVESTOCK FEED, GROUND MULCH, AND SOIL RECONDITIONING

ROOTS
- SOIL STABILIZATION, EROSION CONTROL, AND GROUNDWATER RETENTION
- REMOVE NITRATES, PHOSPHATES AND HEAVY METALS CONTAMINANTS
- TOLERANT TO SOILS WITH HIGH AND LOW PH, SALINITY, AND HEAVY METALS
- DROUGHT AND FIRE RESISTANT
- CARBON SEQUESTERING

Proposed Vetiver Grass Soil Stabilization Plan

Water Resources

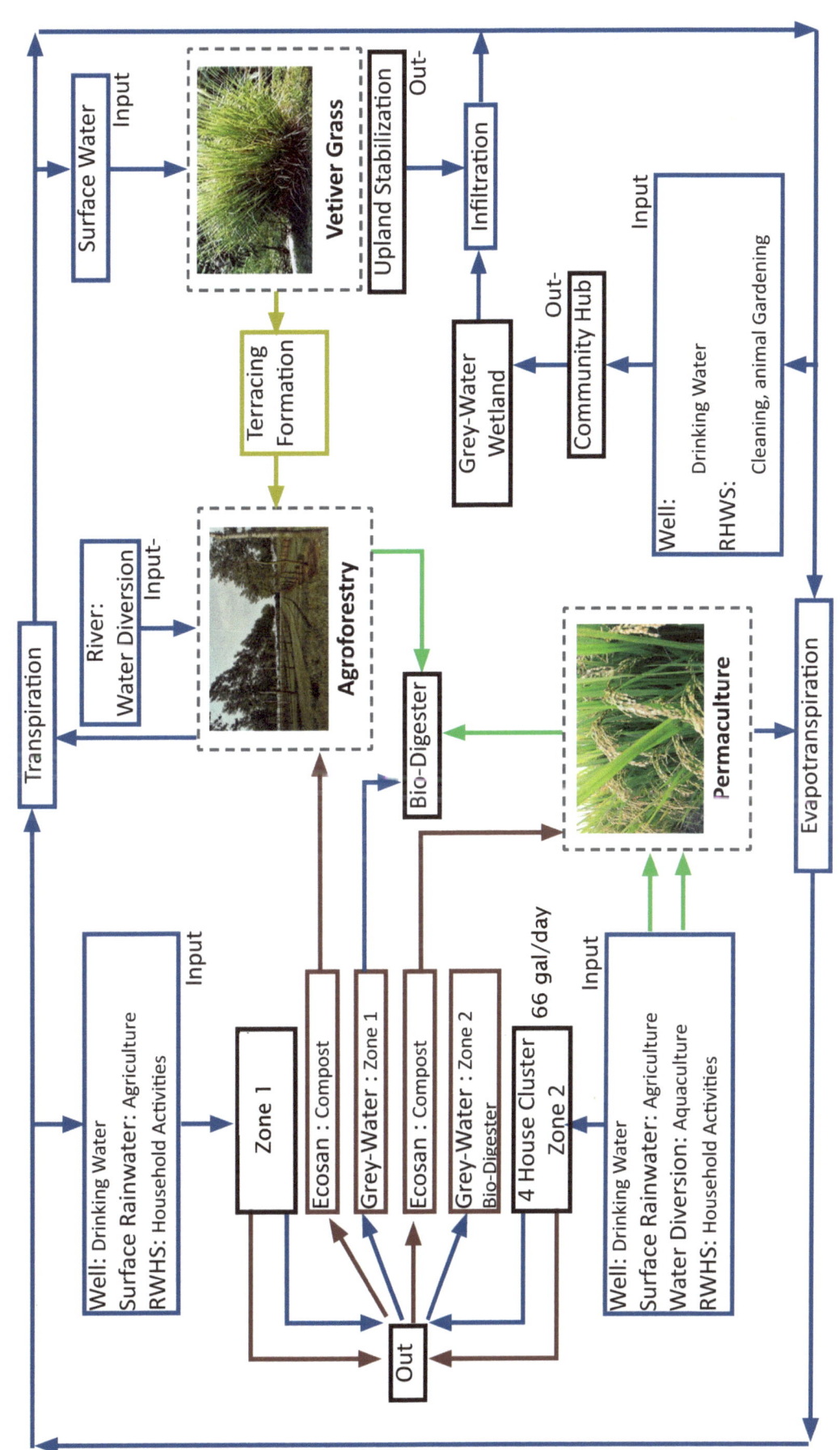

Proposed Water Resources System Flows

Eco-Industrial Park:
Recycling and Concrete Production

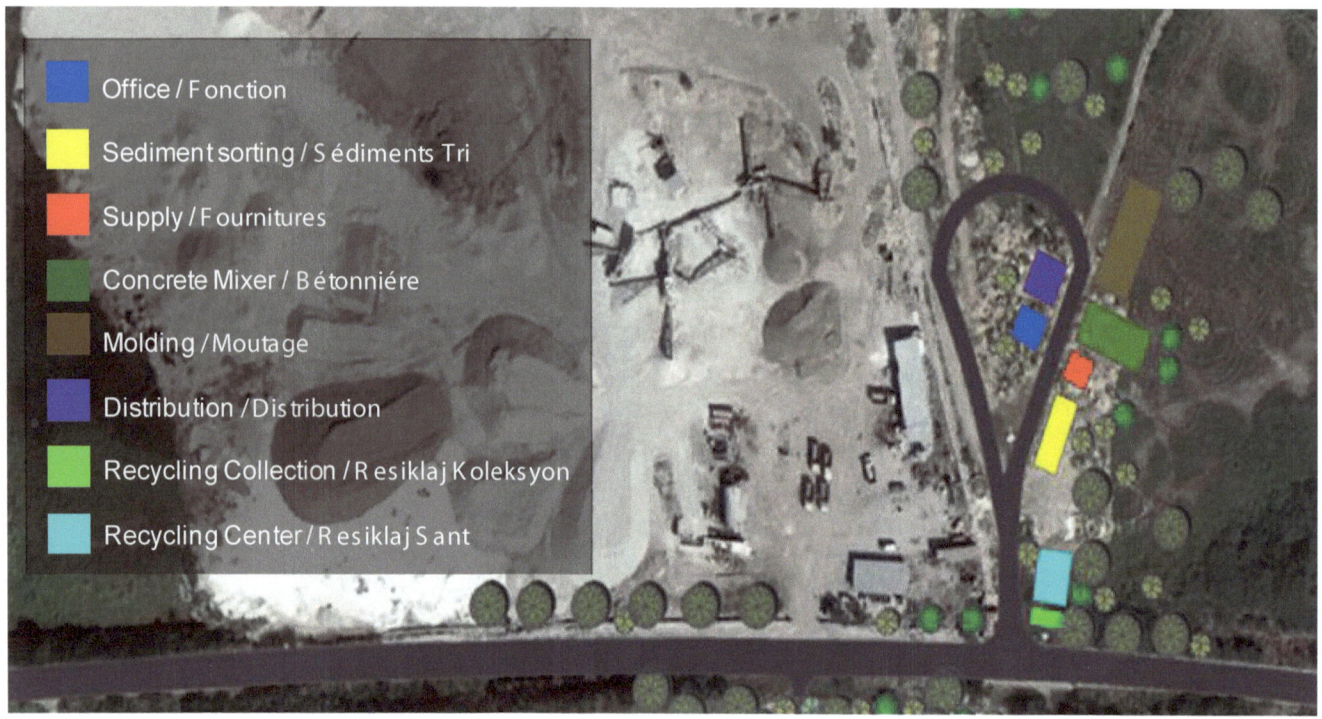

Proposed Concrete Production Site Plan

Problem Statement

Low quality, weak concrete and concrete block, both produced under poorly controlled conditions, contributed significantly to Haiti's high mortality rate during the January 2010 earthquake. Informal production of concrete likely caused ill-proportioned water-cement-aggregate mixes. The high cost of cement in Haiti may have contributed to the wrong ratios. In Fouché, component materials are readily available in the eroded materials found in adjacent river outwashes: sand, stone, and pulverized basalt (a natural pozzolan). Also, the waste accumulating from a mining and gravel-crushing operation in Fouché makes for plentiful concrete aggregate.

Proposed Solution

The combination of the above resources, abundant unskilled local labor, and proximity to urban markets (Grand-Goâve and Leogâne) as well as local need suggest that development of new local concrete industry might be viable. A small, batch-fed concrete mixer, called the "Concrete MD" is a simple machine with labeled buckets that measures materials in correct proportions. It produces about a half cubic meter of concrete per batch. The product, made by Cart-Away Concrete Systems, in Oregon has already been piloted in St. Jacques, Haiti. Initially, a similar operation in Fouché could yield three to four batches per day. Each batch can produce about 2 dozen (10inch by 8inch by 16inch) concrete blocks. Blocks would initially be dedicated for construction of the biodigesters, cisterns and sanitation facilities, as well as the facilities in the Community Hub. Sediments in the Canot River can be accessed during the dry season for use as aggregate, and the plant would operate full time for about 40 weeks a year, with a production estimate of 18,400 blocks with a market value of approximately $ 18,400.

Challenges

The biggest challenge for such a concrete manufacturing enterprise will be paying off the cost of equipment and ongoing cost of cement.

Proposed Phasing

Phase I would entail set up of the initial mixer and market development with a Phase II that would add a second mixer, doubling production.

Eco-Industrial Park

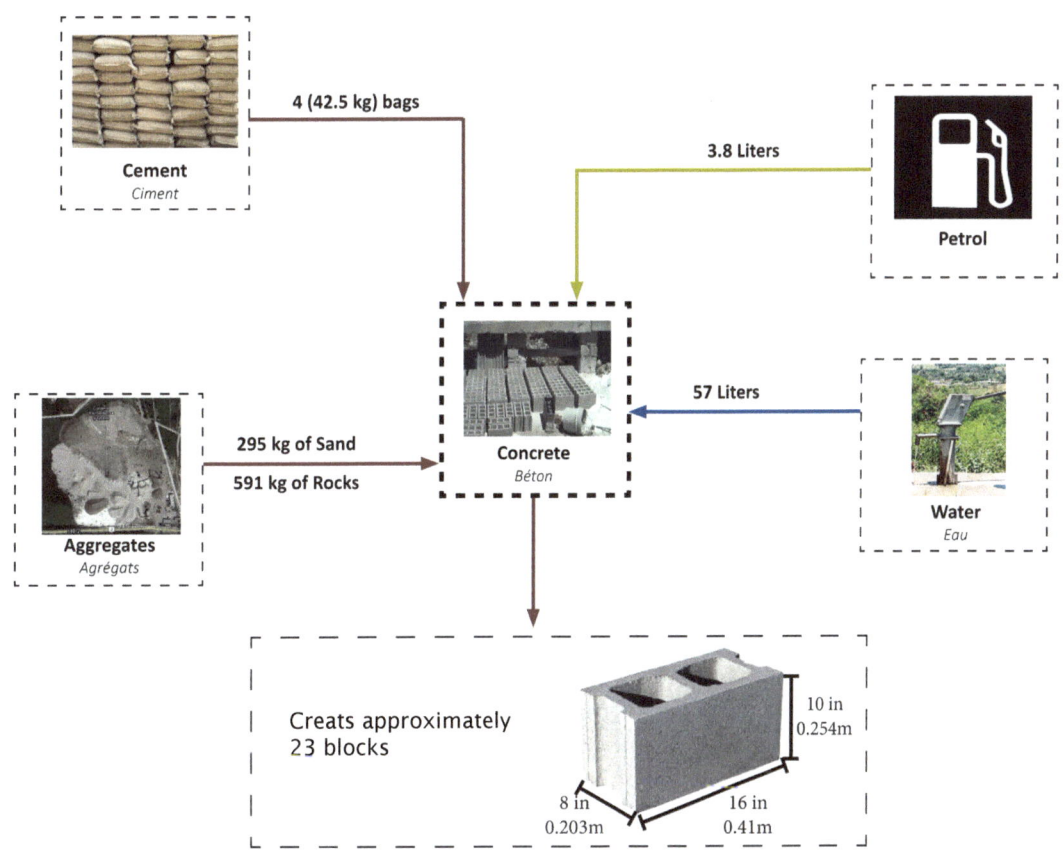

Single Batch Concrete Block Manufacturing Flows

Proposed Concrete Manufacturing and Recycling Center Plan

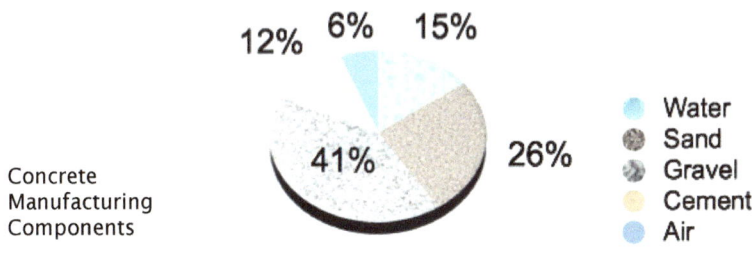

Concrete Manufacturing Components

Legend / Légende
1. Office / Fonction
2. Sediment sorting / Sédiments Tri
3. Supply / Fournitures
4. Concrete Mixer / Bétonniére
5. Molding / Moutage
6. Distribution / Distribution
7. Recycling Collection / Resiklaj Koleksyon
8. Recycling Center / Resiklaj Sant

Infrastructural Ecologies for Fouché, Haiti: Multipurpose, Integrated and Synergistic Systems

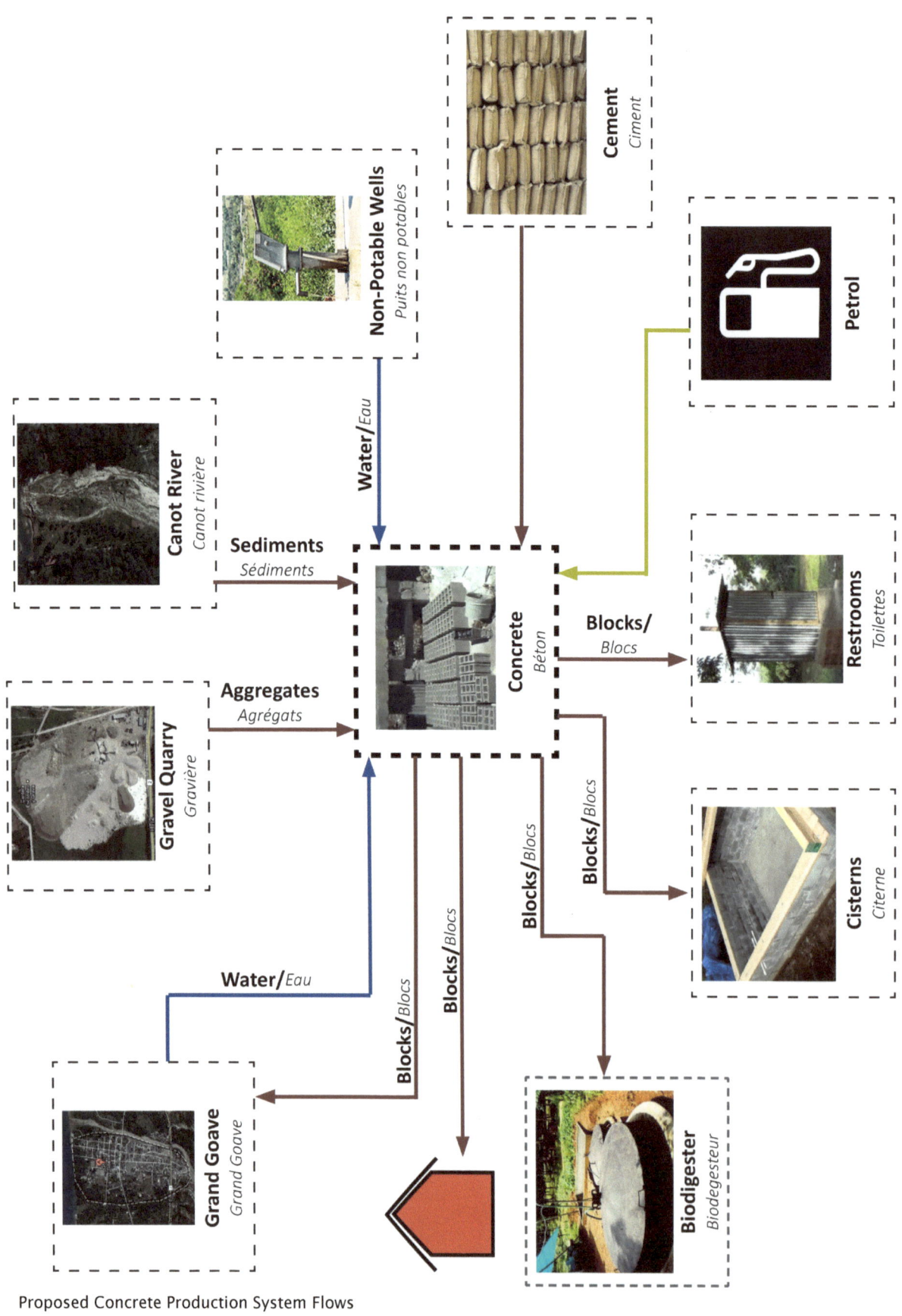

Proposed Concrete Production System Flows

Waste Management

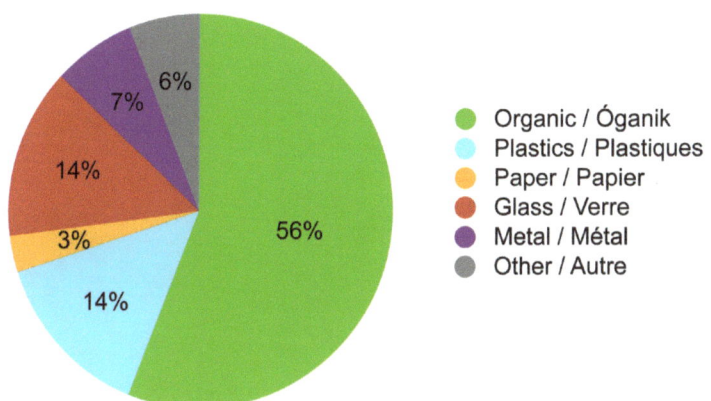

kg/day / *kg/jour*	Haiti	Grand-Goave	Fouché (75% efficiency)			
			%	Zone 1	Zone 2	Total Fouché 1390 inhabitants
Household Organic Waste / *Déchets organiques ménagers*	75%	35%	56	57.12	141.372	198.492
Human Waste / *Déchets humains*	-	-				264
Agriculture Waste / *Les déchets agricoles*	-	-				varies
Grand-Goave Market Waste / *Déchets de marché*		506.8 kg/day*				
Plastics recyclable / *Plastiques recyclables*	7%	20%	7	7.14	17.6715	24.8115
Plastics non-rec			7	7.14	17.6715	24.8115
Paper / *Papier*	5%	1%	3	3.06	7.5735	10.6335
Glass / *Verre*	2%	26%	14	14.28	35.343	49.623
Metal / *Métal*	3%	12%	7	7.14	17.6715	24.8115
Other / *Autre*	8%	4%	6	6.12	15.147	21.267

Existing Waste Composition

Problem statement

The nation of Haiti lacks organized and effective mechanisms to manage its flow of municipal solid waste. Urban collection rates averaging 20% drop to 13% in rural areas. Dump sites are largely uncontrolled. While improvements have been made in the small cities of Petit- Goâve and Grand-Goâve, where garbage is piled in various places in an organized manner, leaving the streets clean, waste bins apparently fill more quickly than can be serviced by collection trucks. Plastic and other waste is also left in the seasonal riverbeds and open canals, clogging drainage and causing threats to human health and flooding risks. Increased awareness and attention to sanitation practices and waste disposal practices is critical.

Proposed Solution

In Fouché, most of the waste is organic matter, and after that, recyclable material. Both streams can be better managed. Several organizations, NGOs and others, have proposed partial solutions to the waste management problem in Haiti, from bottle recycling, to composting toilets for sanitation of human waste, (SOIL). Most organic waste in Fouché is burned or composted. Collection centers for plastic exist, paying 10 cents per pound, and typically cost US $25000 to establish. In Fouché, an estimated 3 to 6 full-time employees in this sector could manage the 17% of MSW that qualifies as recyclable material including plastic bottles, paper and metal. Organic waste management for biogas production can be assisted through domestic sorting

kg/day / *kg/jour*	%	Zone 1	Zone 2	Total Fouche
Household Organic Waste / *Déchets organiques ménagers*	56	57.12	141.372	198.492
Human Waste / *Déchets humains*				264
Agriculture Waste / *Les déchets agricoles*				varies / *varie*
Grand-Goave Market Waste / *Déchets de marché*				253
TOTAL / *TOTALE*				715.492

Waste bins / *Poubelles*
- 56% **Organic waste** / *Dechè òganik*
- 17% **Recyclables** / *Matériau de recyclage*
- 27% **Non-Recyclables** / *Matériau non de recyclage*

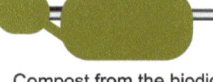

Biogas / *Biogaz*

Zone 1: Biogas collection will go to the community kitchen for cooking.
Zone 2: Gas is used by each household group.

Zone 1: collecte de biogaz ira à la cuisine communautaire pour la cuisson.
Zone 2: Le gaz est utilisé par chaque groupe de ménages.

District Biodigester / *Biodigesteur*

Compost from the biodigester will be returned to farmers in exchange of their agricultural waste.

Compost du biodigesteur sera retourné aux agriculteurs en échange de leurs déchets agricoles.

Organic waste truck / *Camion de déchets organiques*

Zone 1: Residents bring their own organic and agricultural waste by hand or by truck. Waste from Grand-Goave goes to the distric biodigester.
Zone 2: Household groups handle their organic waste.

Zone 1: Les résidents apportent leurs propres déchets organiques et agricoles à la main ou par camion. Déchets de Grand-Goave va au biodigesteur distric.
Zone 2: les groupes de ménages gèrent leurs déchets organiques.

Organic waste bins shall be used to facilitate collection.
Zone 2 will divide the organic waste for biodigester and for animal feed.

Poubelles organiques sont utilisés pour faciliter la collecte.
Zone 2 va diviser les déchets organiques pour biodigesteur et pour l'alimentation animale.

Proposed Organic Material Waste Collection Plan

into binds. Fouché's zone I will receive that zone's organic waste. A system of exchange will be established whereby contributed waste will be exchanged for meals from the community kitchen, cooked using the biogas. House clusters in zone 2 will contribute to and maintain their own biodigesters.

Challenges

Education is necessary to promote a cultural of waste management. Incentivizing sorting and collection may be required.

Proposed Phasing

Before the biodigesters are in place, the must be a learning process in the sorting of waste. Organic waste should go to a composting area, preferably where the biodigester will be. The compost will then be used by farmers and the whole community to build stronger topsoil. Benefits of the compost should encourage people to close the loop on agricultural waste metabolized in the biodigesters. The recycling center will have revenue enough to encourage people to collect plastic bottles on their own.

Waste Management

kg/day / *kg/jour*	%	Zone 1	Zone 2	Total Fouche
Plastics recyclable / *Plastiques recyclables*	7	7.14	17.6715	24.8115
Paper / *Papier*	3	3.06	7.5735	10.6335
Metal / *Métal*	7	7.14	17.6715	24.8115
TOTAL / *TOTALE*				60.2565

Recycling Center / *Centre de Recyclage*

Recycling Collection / *Recyclage collection*

Recycling Collection truck / *Recyclage camion de collecte*

Plastics are sorted and compressed into cubes for shipment.
Facility does not require energy, only manpower.
A recycling center provides between 3 and 6 jobs.

Les plastiques sont triés et compressés en cubes pour l'expédition. Installation ne nécessite pas d'énergie, seulement la main-d'œuvre.
Un centre de recyclage fournit entre 3 et 6 emplois.

Plastics collection, workers are remunerated 10 cents for each pound of material collected. This system is already in use in Haiti.

Plastiques collecte, les travailleurs sont rémunérés 10 cents pour chaque livre de matériel collecté. Ce système est déjà utilisé en Haïti.

For transportation of the materials collected. Other possible solutions are bicycles or motorcycles.

Pour le transport des matériaux collectés. D'autres solutions possibles sont les vélos ou motos.

Sorted waste in different bins: Organics, Recyclables and Non-Recyclables ready for pick-up.

Déchets triés dans des bacs différents: Biologique, recyclables et non recyclables prêts pour le ramassage.

Proposed Recyclable Material Waste Collection Plan

Equipment Location	Equipment/Material	Quantity/Amount	$ Per Unit (L)	$ Per Unit (H)	$ per Unit (average/ or est. for Haiti)	Initial Capital	Maintenance Cost per year
Recycling Center	Building & Equipment	1				$25,000.00	-
	Truck*	1				$18,000.00	-

Energy Elements	Hours Daily Operation	Capacity (w)	Capacity per day (kwh)
Recycling Center	No energy is necessary for operation		

Staff Position/Role	People	Daily Rate	Days	Total
Recycling	3	$5.00	260	$3,900.00

Position/Role	People	Hourly Rate	Pounds / day	Total year
Collectors	informal job	10 cents / pound	52.8	$1,927.20

Overview
Year 1 - Implementation
Equipment $43,000
Energy   0
Staff $5,827.20

Year 2-5-O & M
Equipment   0
Energy   0
Staff $23,308.80

Total (year 1-5) $72,136

*Truck for the Eco-Industrial Park

Proposed Recycling Center Cost Estimates

Infrastructural Ecologies for Fouché, Haiti: Multipurpose, Integrated and Synergistic Systems

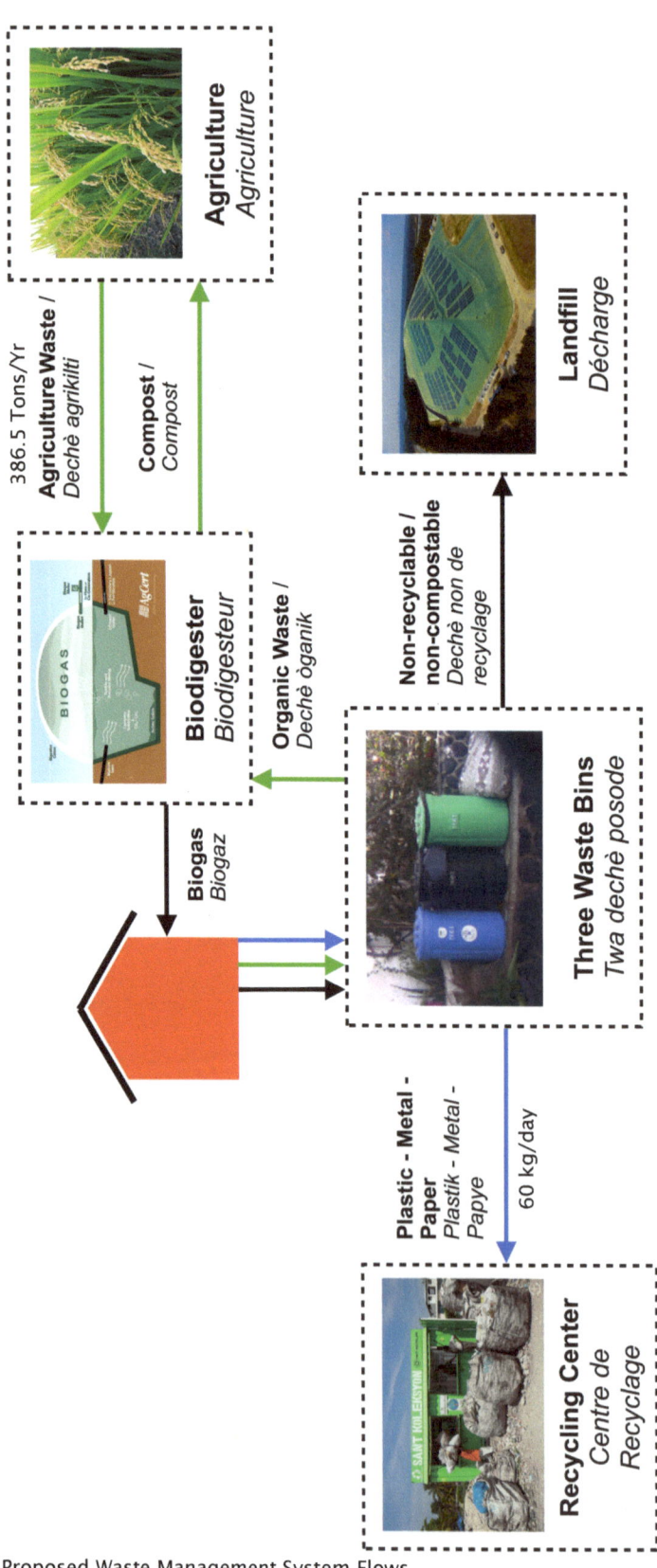

Proposed Waste Management System Flows

Heat and Power – Community Hub and Eco-Industrial Park

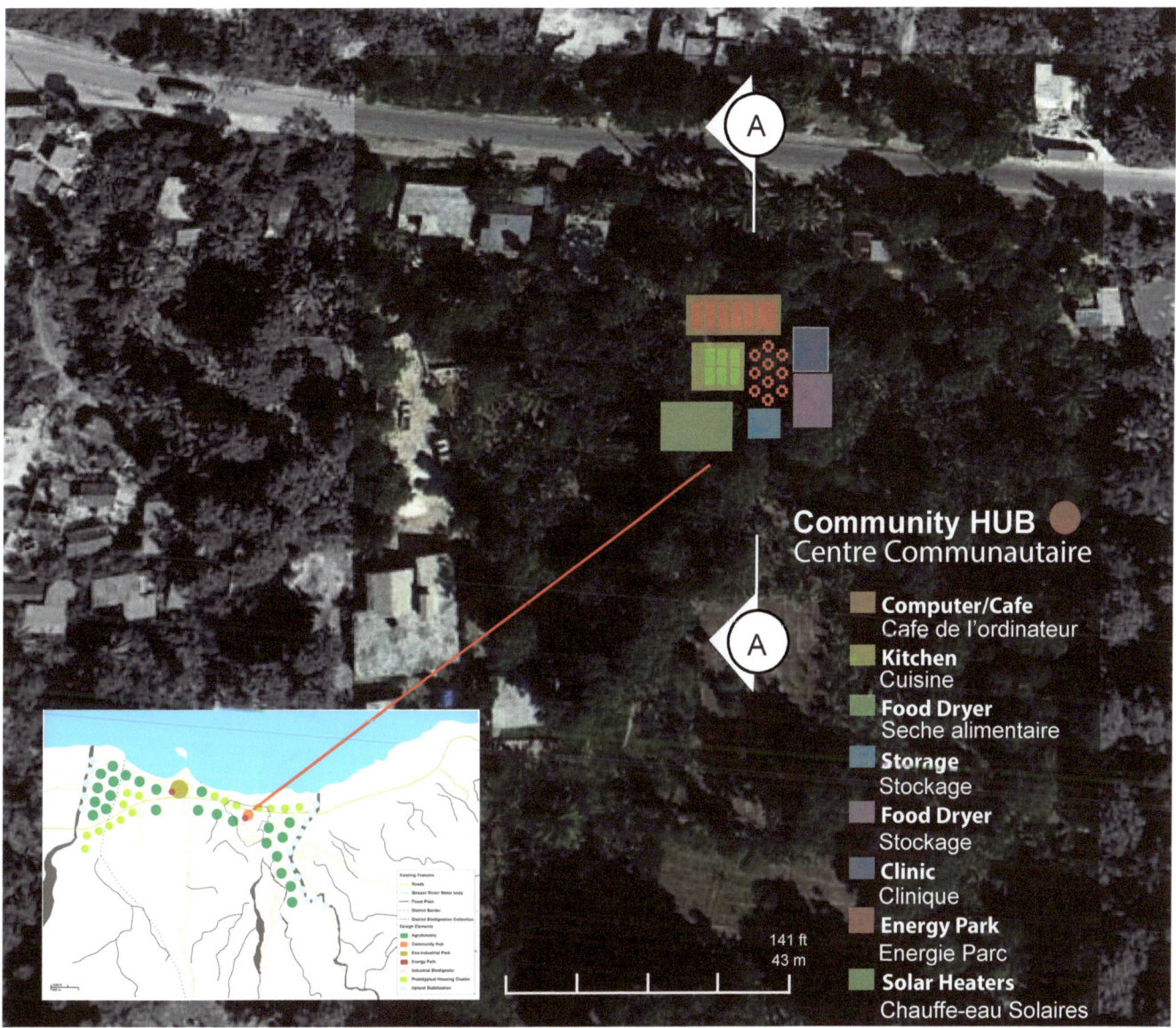

Proposed Energy Park Site Plan

Problem Statement

Haiti endures widespread energy poverty. As the poorest nation in the Americas, its energy resources are limited to imported fossil fuels and sub-optimal hydropower systems for electricity production. Its distribution system is inadequate and unstable. Many users rely on localized diesel generation. In rural areas such as Fouché, there is no access to electricity. Its inhabitants rely, as do some 75% of Haitians, on charcoal for cooking and kerosene for lighting. Such dependency creates health problems due to indoor particulate pollutants. Wood harvesting for charcoal has long contributed to severe deforestation, precipitated dangerous flooding and caused loss of valuable topsoil.

Proposed Solutions

With regard to natural resources, Haiti is excellently suited geographically to utilize relatively simple solar energy systems that could satisfy basic lighting and partial electrification. Wind energy is reliable only in a few locations in the country. However, Haiti could take advantage of its abundant biomass as an energy resource for cooking, lighting and for micro-turbine generated electricity from biogas. Distributed photovoltaic systems, reliable and cost-effective from a life-cycle cost perspective, could adequately support the relatively low demand of a rural community such as Fouché until eventual national electrification. Moreover, inexpensive, solar-powered LED lights can be placed anywhere indoors for domestic use without wiring.

Infrastructural Ecologies for Fouché, Haiti: Multipurpose, Integrated and Synergistic Systems

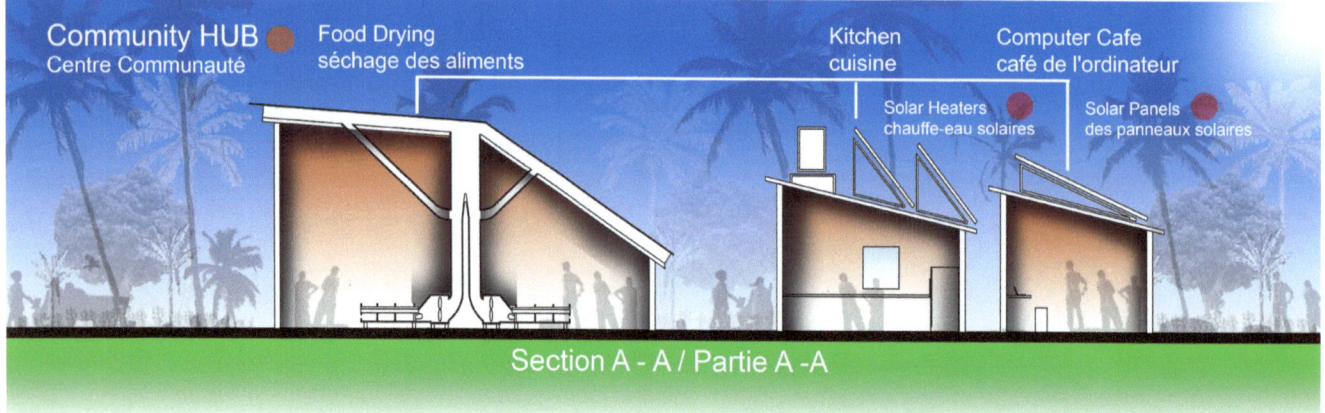

Proposed Community Hub Energy Park Cross-Sectional View

Proposed Community Hub Graphic Representation

The proposed roof-top mounted system for Fouché consists of solar panels (tilted towards south at 15 degrees), along with battery and control units designed to serve the modest electrical needs of the "community hub:" a kitchen, laundry and food drying rooms, a "computer café," and a small clinic. Calculations showed the need for 28 square meters of solar panels, based upon NASA Surface Meteorology statistics. A smaller system is proposed for the currently minimal loads at the eco-industrial park. Hot water needs of the community hub are served by a readily hand-made recycled plastic bottle solar water heater (developed for impoverished areas in Brazil) based on passive thermo-syphonic water circulation.

Challenges

Challenges include the cost of solar panel acquisition and installation, the scarcity of trained electricians for proper panel installation and maintenance. Solar panel theft is another concern and requires some kind of security.

Proposed Phasing

Once the facilities are constructed, the solar hot water system and the PV panels, battery and controls can be added. Implementation of the latter must utilize a trained labor force.

Heat and Power

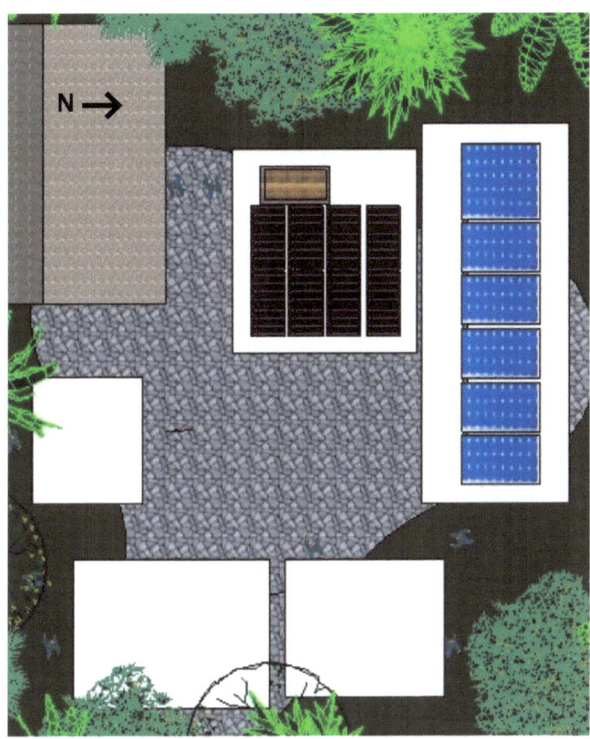

Proposed Community Hub Energy Park Plan

Proposed Community Hub PV Panels Graphic Representation

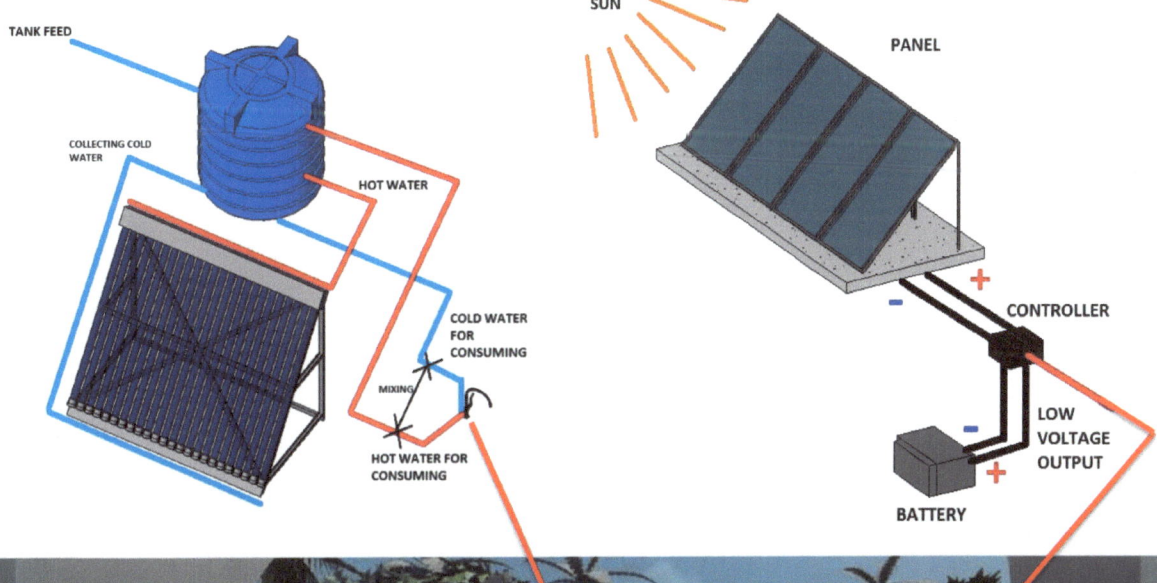

Proposed Community Hub Energy Park Flows Graphic Representation

31

Infrastructural Ecologies for Fouché: Multipurpose, Integrated and Synergistic Systems

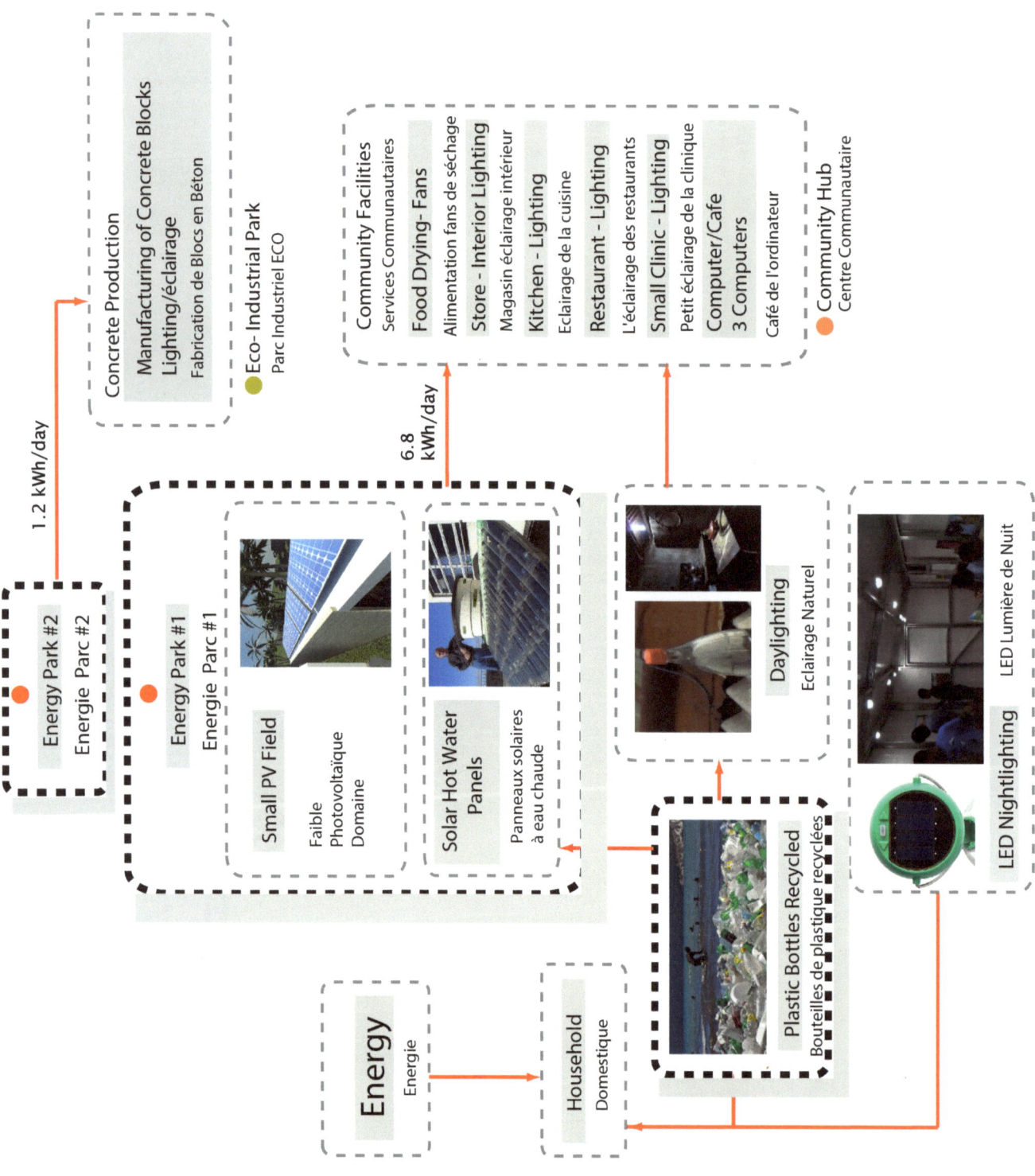

Proposed Energy Park System Flows

Community Hub

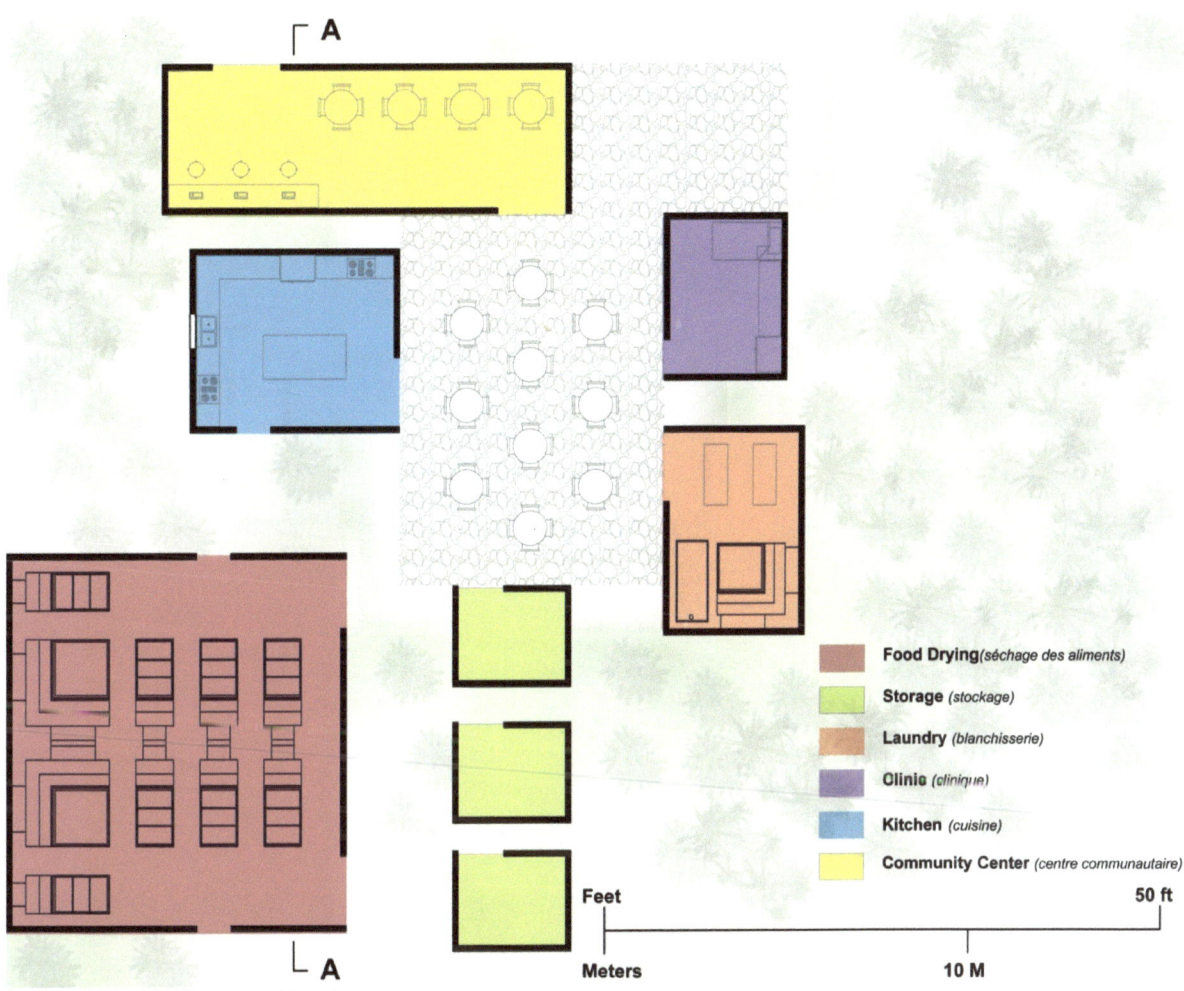

Proposed Community Hub Site Plan

Problem statement

Throughout Haiti's rural regions, agriculture dominates the economy. Few commercial or industrial enterprises exist in the Les Palmes region, itself given largely to farming. As elsewhere in the area, Fouché experiences widespread under-employment. Job creation therefore is a leading driver of this project. Another perceived problem is the diffuse nature of the settlement, with buildings and homes strung out along National Route 2, effectively lacking a village center.

Proposed Solution

The notion of a cooperative "community hub" evolved as a solution to both problems. The hub will be an economic enterprise satisfying several community needs, becoming an effective social center for Fouché, and creating alternative local employment related to the settlement's agricultural production. The hub is comprised of several buildings arranged around a common courtyard with outdoor seating. These include a clinic, a community kitchen (using biogas cooktops) and associated "computer café", a laundry facility, and a solar-powered food drying facility with associated storage, for a total of 11 jobs. The food-dryer would support preservation of fruits, nuts and vegetables for sale. The restaurant is sized to serve 200 meals a day to locals and visitors. The community biodigester is sited nearby to easily receive and process food waste.

Infrastructural Ecologies for Fouché, Haiti: Multipurpose, Integrated and Synergistic Systems

Proposed Solar Food Drying and Community Hub, Cross-Sectional View

Food	Preparation	Dryness Test
Pineapple, Mango, Papaya, Breadfruit	Wash, core, and peel. Cut into 1/4" slices or rings	Leathery with no moisture when cut
Soursop	Wash, halve and pit. "Pop" backs.	Leathery and pliable No moisture when cut
Bananas, Plantain	Peel, slice in thin rounds	Brittle
Almonds, Cashews	Sort, wash, and remove shell.	Brittle and hard
Beans, Vanilla	Wash, chop into small pieces. Blanch 4 minutes	Brittle
Corn	Husk, trim, cut off cob.	Dry and brittle
Sugarcane, coconut	Remove outer skin, then chop.	Brittle
Sorghum, Sisal	Trim	Dry and Brittle

*Food Drying typically takes 3-7 days to be fully dried depending on Humidity Factor

Food Drying Guidelines

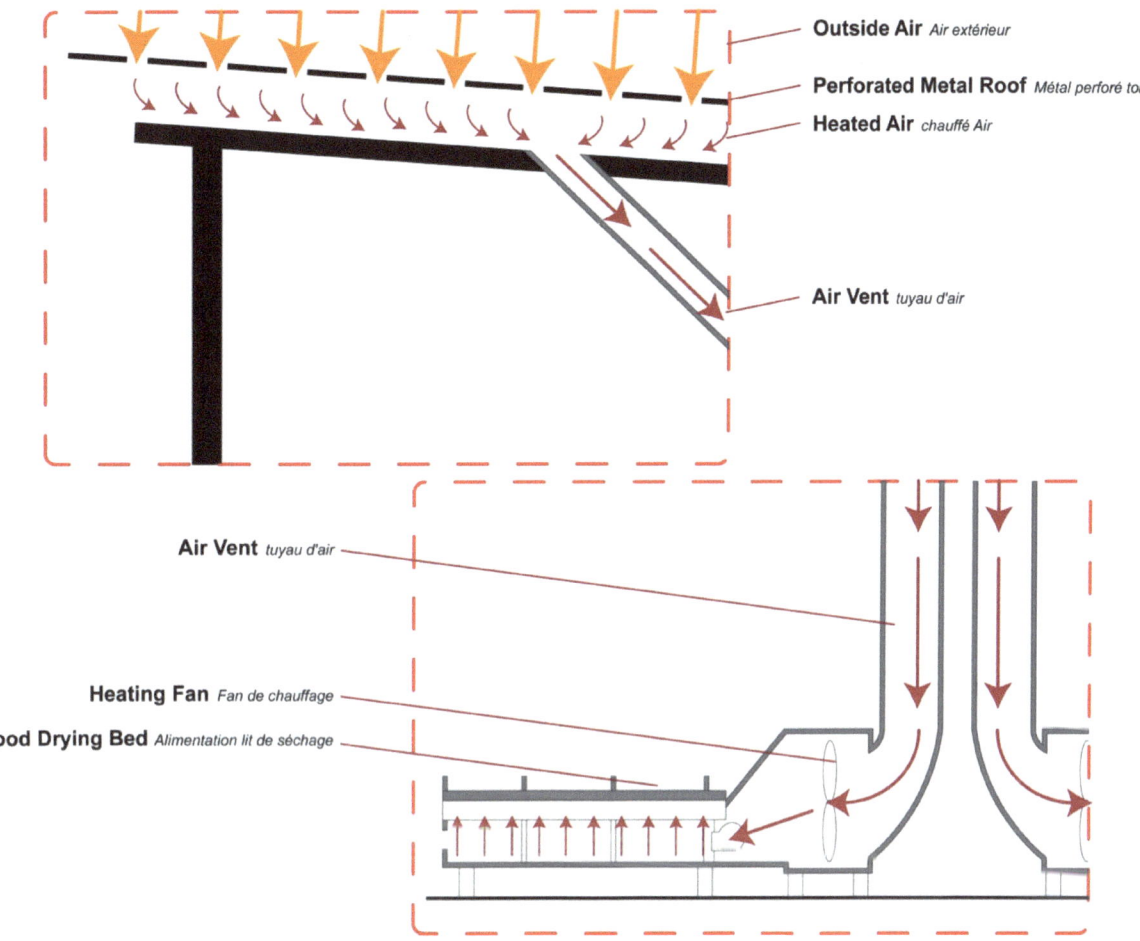

Proposed Community Hub Food Drying Energy Flows Representation

Challenges

Although the buildings themselves could be erected using local materials and labor, the cost to furnish and equip the several enterprises will require additional funds. All-in, it is expected to cost approximately $115,000. Part-time staffing of the clinic would also be an ongoing expense.

Proposed Phasing

The proposed order of construction for the community hub elements are: 1) solar food drying and storage; 2) community kitchen/ computer café; 3) clinic; 4) laundry facility. Funding will be generated by the solar food drying facility to help pay for the other elements of the Hub.

Sector	Initial Capital	Cost per year for Maintenance	Short-term Jobs	Long-term Jobs
A **District Biodigester** *district biodigesteur* *Modular Biodigester costs not included	$ 56,000.00	$ 8,000.00	10	5
B **Community Hub** *hub communauté*	$ 118,500.00	$ 9,600.00	20	12
C **Residential Zone** *zone résidentielle*	$ 70,400.00	$ 5,500.00	15	8
D **Eco-Industrial Park** *Parc éco-industrielle*	$ 76,300.00	$ 15,600.00	5	5
E **Energy Generation** *production d'énergie*	$ 31,600.00	$ 3,500.00	5	2
F **Water Infrastructure** *Infrastructures de l'eau*	$ 40,200.00	$ 11,100.00	20	10
G **Agroforestry** *agroforesterie*	$ 42,300.00	$ 9,700.00	15	28
Total	**$ 491,300.00**	**$ 71,000.00**	**100**	**75**

Proposed Project Cost Calculations Summary

Overall Program Cost

The estimation of project cost was developed by each sector for both the initial implementation and the yearly cost of operations and maintenance. A flat $5 a day rate was used to estimate the cost of local labor in all instances. Equipment was calculated using current prices, in Haiti where possible. The Operations and Maintenance cost comprises an equipment cost that represents 1.0% of the value of the building per year and 5.0% for equipment, as well as the cost of employing local labor. The total cost for initial implementation was calculated as $491,300. Note: distributed biodigester costs not included in this initial estimate as the number of units is not known. The largest costs are in the construction of the community hub and the eco-industrial park. Many of the other sector costs are less because they rely on local labor and community involvement. For example, the change in agriculture to develop the forest garden is reliant on local labor, which will benefit for many years in the future from the additional profits of this system. Similarly, the water infrastructure estimate is dependent on utilizing local people assisted by paid individuals to construct cisterns and toilets for their home. This was done both to maximize funds and to build the community investment into the long-term operation of this infrastructure. The expected employment during the initial construction and implementation is 100 people. This does not include all the people that will need to be involved in the project; 75 long-term jobs are created, including 28 in the agriculture sector due to the expected increases in crop value and productivity.

Attributions

2	Haiti Location Map	Miriam Ward
4	Fouché Existing Conditions	Miriam Ward
5	Proposed Infrastructural Ecology	Hillary Brown, Miriam Ward
6	Fouché Proposed Conditions	Miriam Ward, William Valdez
7	Fouché Area Zoning	Alvaro Munoz Hansen, Arthur Getman
8	Proposed Community Hub Site Plan	William Valdez, Michael McPartland
8	Proposed Community Hub Cross Section	Alvaro Munoz Hansen
9	Community Scale Biodigester	ShanShan Lee
10	Prototype of 5-House Cluster Resource Flows	Alvaro Munoz Hansen
10	Five-House Cluster, Cross Sectional View	Alvaro Munoz Hansen
11	Five-House Cluster Plan	Alvaro Munoz Hansen, Arthur Getman
11	Proposed Modular Biodigester System Flows	Alvaro Munoz Hansen, Shanshan Lee
12	Area-wide Community Biodigester System Flows	Alvaro Munoz Hansen, Shanshan Lee
13	Fouché Agricultural Zone, Existing Conditions	Arthur Getman, Tyler Partato
13	Rainforest Garden Sample Plot	Arthur Getman, Tyler Partato
14	Proposed Edible Rainforest Garden	Arthur Getman, Tyler Partato
15	Edible Rainforest Garden 3 Cross-Sectional Views	Arthur Getman, Tyler Partato
16	Proposed Agricultural Zone System Flows	Arthur Getman, Tyler Partato
17	Fouché Proposed Water Resources	Diana Bedoya, Jose Germosen
18	Rainwater Catchment Cistern Proposal	Diana Bedoya, Jose Germosen
19	Vertical Flow Wetlands	Diana Bedoya, Jose Germosen
19	Ecosan Toilet Proposal	Diana Bedoya, Jose Germosen
20	Proposed Runoff Macro-Catchment and Diversion Plan	Diana Bedoya, Jose Germosen
20	Proposed Vetiver Grass Soil Stabilization Plan	Diana Bedoya, Jose Germosen, http://vetiverlatrine.org/about.php
21	Proposed Water Resources System Flows	Diana Bedoya, Jose Germosen
22	Proposed Eco-Industrial Park	Melissa Cheing, Kimberly Urko
23	Single Batch Concrete Block Manufacturing Flows	Kimberly Urko
23	Concrete Manufacturing Components	Kimberly Urko
24	Proposed Concrete Production System Flows	Kimberly Urko
25	Existing Waste Composition	Melissa Cheing
26	Proposed Organic Material Waste Collection Plan	Melissa Cheing
27	Proposed Recyclable Material Waste Collection Plan	Melissa Cheing
27	Proposed Recycling Center Cost Estimates	Melissa Cheing
28	Proposed Waste Management System Flows	Melissa Cheing
29	Proposed Energy Park Site Plan	William Valdez
30	Proposed Community Hub Energy Park Cross-Sectional	William Valdez, Maria Bellini Whalen
31	Proposed Community Hub Energy Park Plan, PV Panels	Maria Bellini Whalen
31	Energy Park Flows Graphic Representation	Maria Bellini Whalen
32	Proposed Energy Park System Flows	William Valdez, Maria Bellini Whalen
33	Proposed Community Hub Site Plan	Michael McPartland
34	Solar Food Drying and Community Hub & Guidelines	Michael McPartland
35	Food Drying Energy Flows Representation	Michael McPartland
36	Proposed Project Cost Calculations Summary	Miriam Ward

www.ingramcontent.com/pod-product-compliance
Lightning Source LLC
Chambersburg PA
CBHW041257180526
45172CB00003B/884